Pavel Shibaev

Introduction to professional activities

AF533266

Pavel Shibaev

Introduction to professional activities

Workshop

ScienciaScripts

Imprint
Any brand names and product names mentioned in this book are subject to trademark, brand or patent protection and are trademarks or registered trademarks of their respective holders. The use of brand names, product names, common names, trade names, product descriptions etc. even without a particular marking in this work is in no way to be construed to mean that such names may be regarded as unrestricted in respect of trademark and brand protection legislation and could thus be used by anyone.

Cover image: www.ingimage.com

This book is a translation from the original published under ISBN 978-620-7-80864-9.

Publisher:
Sciencia Scripts
is a trademark of
Dodo Books Indian Ocean Ltd. and OmniScriptum S.R.L publishing group

120 High Road, East Finchley, London, N2 9ED, United Kingdom
Str. Armeneasca 28/1, office 1, Chisinau MD-2012, Republic of Moldova, Europe
Printed at: see last page
ISBN: 978-620-7-90929-2

Copyright © Pavel Shibaev
Copyright © 2024 Dodo Books Indian Ocean Ltd. and OmniScriptum S.R.L publishing group

Reviewers:
D. in Technical Sciences, Associate Professor at Vologda State University,
A.A. Sinitsyn, Director of the Research Center "Problems of Modern Techno-Environment";
Candidate of Technical Sciences, Senior Lecturer of FGBOU VO "KNITU" *A.A. Davletshin*

CONTENTS.

INTRODUCTION

The purpose of training is to form students' knowledge of the basics of materials science and areas of activity of a full-time bachelor in accordance with the profile "Materials Science and Technology of Materials", including consideration of the development of materials science in the historical context, while revealing, through the specificity of the object and subject of study the individuality of this scientific and educational discipline, objects, areas and types of professional activities of the graduate.

Tasks of the discipline (module)

The main objectives of the discipline are:

1. - familiarization with the history and main stages of development of materials science;
2. - familiarization with modern theories of structure of non-metallic and metallic materials, methods of studying their structure and properties;
3. - familiarization with basic concepts, terms and definitions in materials science;
4. - familiarization with the main groups of traditional and modern materials of various functional purposes used in modern technology;
5. - Familiarization with the main parameters used to evaluate material properties;
6. - familiarization with the main directions of regulating the structure and complex of technical properties of materials to create promising materials with specified properties.

Content of sections of the discipline (module)

Section 1. Introduction to the specialty and stages of materials science development. Object, subject and basic concepts of materials science and materials technology.

Topic 1.1. General characteristics of the object and subject of materials science. System of basic and auxiliary concepts of materials science. The main tendencies of expanding the nomenclature of materials used in practice (metals, organic polymers, ceramics, composite materials, etc.) and problems (theoretical, applied and raw materials) of modern materials science. Purpose and objectives of the discipline "Introduction to Professional Activity".

Topic 1.2 Features of the profession of a materials scientist.

The field, objects, types and tasks of professional activity of a graduate (in research, design, production and technological, organizational and managerial activities).

General characteristic of the historical development of materials science. Materials in the life support of mankind. Brief characterization of the stages of materials science development.

Section 2: General characterization of structure (structure) of materials.

Topic 2.1. General characteristics of structural levels (micro-, meso- and macro) of materials organization. Specificity in the structure of metallic and non-metallic materials.

Section 3: Properties of materials and their varieties.

Topic 3.1. General characterization of materials properties and their classification. Influence of structure on material properties. Difference in characteristic properties of three main groups of substances and materials (metallic, polymeric and ceramic).

Section 4: Classification of materials, areas of their application.

Topic 4.1. Classification of materials by various features. Characterization of the main areas of materials application in the real sector of economy.

Section 5: Development of Materials Science in Historical Retrospect.

Topic 5.1. Chronology of material science achievements in different historical periods. Contribution of individual personalities to the development of materials science.

Topic 5.2. Achievements of domestic materials scientists and their contribution to the world scientific and technical progress.

Topic 5.3. History of evolution of expansion of the nomenclature of materials practically used by mankind. Volumes of world production of main types of materials in the XXI century.

TOPIC 1. EDUCATIONAL PROGRAM IN THE DIRECTION 22.03.01 "MATERIALS SCIENCE AND MATERIALS TECHNOLOGY"

Objective. To study and analyze the educational program (EP) and annotations to the working programs of disciplines and practices in the direction 22.03.01 "Materials Science and Technology of Materials".

Description of the OP

1.1 Rationale for the development of the OP

The educational program defines the requirements for the implementation of educational activities in the direction of Bachelor's degree 22.03.01 "Materials Science and Technology of Materials". OP is an innovative (advanced) educational program of higher education (Bachelor's degree level), corresponding to the priority direction of the Strategy of scientific and technological development of Russia (Decree of the President of the Russian Federation from 01.12.2016 N 642 "Strategies of scientific and technological development of the Russian Federation"). When developing the program, the global scientific and technological trends in the development of digital and intelligent production technologies, robotic systems, new materials are taken into account.

The educational program is focused on the formation of competencies that will increase the competitiveness of program graduates in the labor market. Graduates of the program are prepared for research and technological tasks of professional activity in accordance with the direction and focus of training.

The Program of Higher Education is a system of documents developed and approved by a higher education institution taking into account the needs of the regional labor market, the requirements of federal executive authorities and relevant industry requirements on the basis of the federal state educational standard of higher education in the relevant field of study.

OP BO regulates the goals, expected results, content, conditions and technologies of the educational process, evaluation of the quality of graduate training in this direction of training and includes: curriculum, working programs of academic disciplines (modules) and practices and other materials that ensure the quality of training of students, as well as the calendar academic schedule and methodological materials that ensure the implementation of appropriate educational technology, a working program of education and calendar plan of educational work

1.2 Regulatory documents for the development of the EP of higher education in the direction of training

Realization of educational activity in the direction (specialty) is carried out on the basis of the requirements of the following basic documents:

- Federal Law of 29.12.2012 No. 273-FZ "On Education in the Russian Federation";

- Federal State Educational Standard of Higher Education in the direction of training 22.03.01 "Materials Science and Technology of Materials", approved by the order of the Ministry of Education and Science of the Russian Federation "June 02", 2020 № 701.

-The Order of the Ministry of Education and Science of the Russian Federation from 05.04.2017 № 301 "On Approval of the Order of organization and implementation of educational activities under educational programs of higher education - bachelor's degree programs, specialist programs, master's degree programs";

-GOST 7.32-2001 Interstate standard. System of standards on information, librarianship and publishing. Report on research work. Structure and rules of execution;

-GOST ISO 9000-2011 Interstate standard. Quality management systems. Basic provisions and vocabulary;

-GOST ISO 9001-2011 Interstate standard. Quality management systems. Requirements;

-Constitution of KNITU-KAI;

-MI. 4.2.3-01-2014 General requirements for the content, design and management of the regulation on activities (regulations on the implementation of processes) of KNITU-KAI;

- P.8.1-01-2017 Procedure for organization and implementation of educational activities under educational programs of higher education - bachelor's degree programs, specialist programs, master's degree programs.

- P.7.3-2018 Regulations on the procedure of development and approval of educational programs of higher education - bachelor's degree programs, specialist programs, master's degree programs.

- P-9.1_8.3.5-05-2019 Regulation on the GIA for the Bachelor's, Specialist's and Master's degree programs.

1.3.General characterization of the OP

Training Direction:

22.03.01 "Materials science and technologies of materials"

The orientation (profiles) of the educational program:

"Materials science and technology of new materials"

Qualification (degree): ***Bachelor***

Form of training ***full-time***

Normative period of mastering: ***4 years***

Labor intensity of the program *240 credits: 8968 hours.*

Entry Requirements:

the applicant must have a state sample document on secondary general education or a document on secondary vocational education, or a document on higher education and qualification.

1.4 Mission, Goals and Objectives of the PGCE

The purpose of the Bachelor's program in the direction of training 22.03.01 "Materials Science and Technology of Materials": development of personal qualities in students, as well as the formation of universal, general professional and professional competencies in accordance with the requirements of the Federal State Standard of Higher Education in the direction of training 22.03.01 "Materials Science and Technology of Materials".

The purpose of the OP in the field of personal education is to strengthen morality, development of general cultural needs, creative abilities, responsibility, social adaptation, communication, tolerance, perseverance in achieving goals, endurance and physical culture.

The purpose of the OP in the field of education is to meet the needs of the individual in mastering knowledge in the field of humanities, social, economic, mathematical and natural sciences and professional disciplines, allowing the graduate to work successfully in the relevant field of activity, to possess universal and professional competencies, contributing to his social mobility and demand in the labor market. Achievement of the goal is ensured by methodological, organizational, personnel, material and technical components of the educational process that meets the requirements of the world level of education in this subject area.

The mission of the OP is to form the intellectual elite, creating the most favorable environment for the development of education, research and innovation, for the formation of new generations of scientific, scientific-pedagogical and engineering-technical personnel capable of leading and keeping the Russian industry in the orbit of the world market.

Bachelors training in the direction 22.03.01 "Materials Science and Technology of Materials" is carried out with a broad coverage and deep study of the complex of problems associated with the creation and application of various metallic and non-metallic materials and coatings, as well as technologies for their production and processing. This direction, which can make the greatest contribution to ensuring the security of the country, accelerating economic growth, increasing the competitiveness of the country through the development of the technological base of the economy and knowledge-intensive industries, is included in the List of priority directions of development of science, technology and engineering of the Russian Federation.

The demand for graduates of this direction is connected with the fact that in the conditions of modern digital production and a new technological mode it is necessary to have specialists who use in practice modern knowledge of materials sciences, the influence of macro-, micro- and nano-scale on the properties of materials, interaction with the environment, as well as traditional and new technological methods of synthesis and analysis of advanced materials, processes of obtaining and processing of products for various purposes.

That is why the department of "Materials Science, Welding and Industrial Safety" carries out training of bachelors in the direction 22.03.01 "Materials Science and Technology of Materials", the most demanded for the enterprises of mechanical engineering, medicine, oil and gas, energy and aerospace industries of our and other regions of Russia and neighboring countries. Our students get theoretical and practical knowledge and skills of using methods of research, analysis and diagnostics of materials properties, selection of rational technologies of their obtaining, processing and modification.

Materials science plays an important role in the development of scientific and technological progress. In any branch of national economy. Materials science is able to cover a wide range of human activities, which makes this direction relevant and in demand in our time of rapid development of new technologies and materials.

Obtaining, development and analysis of new materials, methods of their processing are the basis of modern production and largely determine the level of development of scientific, technical and economic potential of the state.

Materials science has been developing especially intensively in recent decades. This is explained by the need for new materials for space exploration, medicine, development of electronics, nuclear power engineering. This requires the inclusion of almost all elements of the periodic system in the number of industrial materials.

The solution of the most important technical tasks related to economical consumption of materials, weight reduction of machines and devices largely depends on the development of materials science. The continuous process of creating new materials for modern technology enriches the science of materials.

Today the branch of materials science is a high-tech sphere of activity. Materials science is included in the list of priority areas of development in all developed countries of the world, and in itself is one of the most demanded industries.

The Republic of Tatarstan is a major center of general and special machine building, instrument making, aircraft building, helicopter building, construction and chemical industry. These industries are among the leading sectors of the economy of Kazan and the Republic of Tatarstan, making a significant contribution to the employment of the economically active population. Industrial production is one of the most dynamically developing spheres of economic activity in the Republic of

Tatarstan. A large number of relevant enterprises, design and research organizations of national and international importance are located here. Manufacturing of products by methods of pressure, casting and welding are the basis of blank production of any enterprise. At the same time, the design of rational and competitive products, organization of their production is impossible without a sufficient level of knowledge in the field of materials science, which is the most important indicator of education of a modern graduate specialist.

All this requires the training of appropriate specialists in materials science and determines the need for highly qualified personnel for enterprises, organizations and research institutes.

Professional activity of graduates in the direction of training 22.03.01 "Materials Science and Technology of Materials" is associated with technological processes of obtaining, processing and recycling of modern materials, the study of their chemical and morphological composition, phase state, certification of materials and coatings, with technological processes of their obtaining, processing and quality control. Under the guidance of teachers and mentors of KNITU-KAI bachelors of training direction 22.03.01 "Materials Science and Technology of Materials" fully master the scientific specialty and become professionals in their field of activity.

Employees of the Department of Materials Science, Welding and Industrial Safety have developed a large number of unique textbooks and manuals, taking into account domestic and international experience in the field of processing and recycling of modern materials.

The main consumers of bachelors of training direction 22.03.01 "Materials Science and Technology of Materials" are large modern enterprises of various industries, such as AK "Transneft", "Gazpromtransgaz Kazan", JSC "KMPO", KAZ named after S.P. Gorbunov (branch of PJSC "Tupolev"), FSUE "Tochmash", JSC "KMIZ", JSC "Radiopribor". Gorbunov (branch of PJSC "Tupolev"), "KVZ - Helicopters of Russia", FSUE "Tochmash", JSC "KMIZ", JSC "Radiopribor", "KAMAZ", JSC "Kazankompressormash", JSC "PO "Zavod named after Sergo", JSC "Zelenodolsk plant named after A.M.

Gorky", automotive service enterprises, diagnostic centers, non-destructive and destructive testing, etc.

The demand for engineering personnel in the specialties of training direction 22.03.01 "Materials Science and Technology of Materials" is confirmed by the applications of JSC "KMPO", S.P. Gorbunov KAZ (branch of PJSC "Tupolev"), JSC "Kazankompressormash", "KVZ - Helicopters of Russia" for target training.

Section 2: Characteristics of professional activity of a graduate in the direction of training

2.1 Area and spheres of professional activity of the graduate

The area of professional activity of graduates who have mastered the Bachelor's degree program includes:

- development, research, modification and use (processing, operation and utilization) of metallic and non-metallic materials for various purposes;
- processes of their formation, shape and structure formation, transformation at the stages of production, processing and operation;
- processes of obtaining materials, blanks, semi-finished products, parts and products, as well as their quality management for various fields of engineering and technology (mechanical engineering and instrumentation, aviation and rocket-space engineering, nuclear power engineering, solid-state electronics, nanoindustry, medical equipment, sports and household equipment).

Spheres of professional activity of the graduate, for which the graduates who have mastered the Bachelor's degree program in the direction of training 22.03.01 (Materials science and technology of materials (profile) "Materials science and technology of new materials") are prepared are correlated with professional standards.

The list of professional standards, generalized labor functions and labor functions relevant to the professional activity of the graduate of the Bachelor's program in the direction of training 22.03.01 (Materials science and technology of materials (profile) "Materials science and technology of new materials"), are given in Table 1.

Table 1. Areas of professional activity and labor functions according to the field of professional activity

List of professional standards	Name of the sphere of professional activity	Generalized labor functions	Labor functions
40.136 Specialist in the development, maintenance and integration of technological processes and productions in the field of materials science and materials technology	Creation and management of integrated technological processes and productions in the field of materials science and technology. **Primary Objective:** Ensuring high efficiency of production with optimal technical and economic indicators	Development, maintenance and integration of standard technological processes in the field of materials science and materials technology	A/01.6 Development of typical technological processes in the field of materials science and materials technology A/03.6 Support of typical technological processes in the field of materials science and materials technology
40.085 Thermal Production Quality Control Specialist	Heat treatment **Primary Objective:** Ensuring the specified quality of thermal production products with the established	Technological control and performance of operations to assess the quality of thermal	A/01.5 Control of surface material and/or volume characteristics of parts after heat treatment

	technical and economic indicators	production products	A/02.5 Uncomplicated investigations supplied by more qualified persons A/03.5 Control of compliance with technological discipline A/05.5 Preparation of samples and structure analysis for compliance with normative documentation

2.2 Objects of graduate's professional activity

The objects of professional activity of bachelors in the direction 22.03.01 "Materials science and technology of materials" are:

- basic types of modern structural and functional inorganic (metallic and non-metallic) and organic (polymeric and carbon) materials; composite materials; superhard materials; films and coatings;

- methods and means of testing and diagnostics, research and quality control of materials, films and coatings, semi-finished products, blanks, parts and products, all kinds of research, control and testing equipment, analytical equipment, computer software for processing of results and analysis of obtained data, modeling of materials behavior, evaluation and forecasting of their performance characteristics;

- technological processes of production, processing and modification of materials and coatings, parts and products; equipment, technological equipment and fixtures; technological process control systems;

- normative and technical documentation and certification systems of materials and products, technological processes of their production and processing; reporting documentation, records and protocols of the course and results of experiments, documentation on safety and life safety.

2.3 Tasks of the graduate's professional activity

A graduate who has mastered the Bachelor's program in the direction 22.03.01 "Materials Science and Technology of Materials", in accordance with the research and technological types of professional activity should be ready to solve the following professional tasks:

research type of tasks

- BAT 1 - collection of data on existing types and grades of materials, their structure and properties in relation to the solution of the set tasks using databases and literature sources;
- NID 2 - participation in the work of a group of specialists in the performance of experiments and processing of their results on the creation, research and selection of materials, assessment of their technological and service qualities by complex analysis of their structure and properties, physical-mechanical, corrosion and other tests;
- NID 3 - collection of scientific and technical information on the subject of experiments for the preparation of reviews, reports and scientific publications, participation in the preparation of reports on the completed assignment;
- NID 4 - work with normative and technical documentation in the system of certification of materials and products, technological processes of their obtaining and processing, reporting documentation, records and protocols of the course and results of the experiment, documentation on safety and life safety;
- NID 5 - participation in the work of a group of specialists in the development of technological processes of production, processing and modification of materials and coatings, parts and products, technological process control systems;

- NID 6 - record keeping, execution of project and working technical documentation, preparation of records and protocols at production sites in accordance with the requirements of regulatory documentation;

Technology type of tasks:

- TD 1 - participation in obtaining and use (processing, operation and utilization) of materials for various purposes, design of high-tech processes at the stage of pilot testing and implementation;

-TD 2 - Participate in the organization of workplaces in the division, maintenance and diagnostics of measuring instruments and test equipment, control of compliance with quality requirements for measurements and tests, and data processing;

-TD 3 - participation in the development of technical specifications for measurements, tests, research and development works;

-TD 4 - participation in work on standardization, preparation and carrying out of certification of processes, equipment and materials, preparation of documents when creating a quality management system in the organization;

-TD 5 - Design of high technology processes as part of a primary design and technology or research unit;

-TD 6 - development of working technical documentation.

Section 3. Graduate Competencies as a cumulative expected result of education upon completion of the given Program of Graduate Studies

3.1 A graduate must possess the following universal competences (UC)

№	Competence to be formed	Code
1.	Able to search, critically analyze and synthesize information, apply a systematic approach to solve the tasks at hand	UK-1
2.	Able to define the range of tasks within the set goal and choose the best ways to solve them, based on the current legal norms, available resources and limitations	CC -2

3.	Able to carry out social interaction and realize his/her role in a team	CC -3
4.	Able to conduct business communication orally and in writing in the state language of the Russian Federation and foreign language(s)	CC -4
5.	Able to perceive the intercultural diversity of society in socio-historical, ethical and philosophical contexts	CC - 5.
6.	Able to manage his/her time, build and implement a trajectory of self-development based on the principles of lifelong learning	MC- 6.
7.	Able to maintain an adequate level of physical fitness to ensure full social and professional activity	MC - 7
8.	Able to create and maintain safe living conditions in everyday life and in professional activities to preserve the natural environment, to ensure sustainable development of society, including in case of threat and occurrence of emergencies and military conflicts	CC -8
9.	Able to use basic defectological knowledge in social and professional spheres	UK-9
10.	Able to make informed economic decisions in various areas of life activity	UK- 10
11.	Capable of forming an intolerant attitude towards corrupt behavior	UK- 11

3.2 A graduate should possess the following general professional competencies (GPC)

№	Competence to be formed	Code
1	2	3
1.	Able to solve problems of professional activity, applying methods of modeling, mathematical analysis, natural science and general engineering knowledge	OPK- 1
2.	Able to participate in the design of technical objects, systems and technological processes taking into account economic, environmental and social constraints	OPK- 2
3.	Able to participate in the management of professional activities using knowledge in the field of project management	OPK- 3
4.	Able to carry out measurements and observations in the field of professional activity, process and present experimental data	OPK- 4

5.	Able to solve research problems in the implementation of professional activities with the use of modern information technologies and applied hardware and software tools	OPK-5
6.	Able to make sound technical decisions in professional activities, select effective and safe technical means and technologies	OPK-6
7.	Able to analyze, draft and apply technical documentation related to professional activities in accordance with the current regulatory documents in the relevant industry	OPK-7

3.3 The graduate shall possess the following professional competencies (PC)

№	Competence to be formed	Code
Type of professional activity tasks: research		
1.	Able to use knowledge of materials and methods of research, analysis, diagnostics and modeling of properties of substances (materials), physical and chemical processes occurring in materials during their obtaining, processing and modification to solve problems of professional activity	PC-1
2.	Ready to perform standard measurements, conduct tests in the study of materials and products, processes of their production, processing and modification, including for certification purposes	PC-2
Type of professional activity tasks: technological		
1.	Able to use knowledge of the relationship between the composition, structure and properties of modern metallic and non-metallic materials to solve problems of professional activity	PC-3
2.	Capable of selecting metallic and non-metallic materials of machine parts, mechanisms and structures, as well as processing methods depending on their purpose	PK-4
3.	Knows typical technological processes of obtaining, processing, treatment and modification of materials and is able to choose technological equipment for their realization	PK-5
4.	Able to create records, protocols, design and technological documentation in accordance with the	PK-6

	requirements of regulatory documents, including using standard software tools	

Section 6: Normative and methodological support of the system for assessing the quality of mastering the Bachelor's Program of Higher Professional Education

Mastering of the OP, including a separate part or the whole volume of the discipline (module), is accompanied by current progress control and interim attestation of students.

Current progress control provides assessment of the progress of mastering disciplines (modules) and practical training.

Intermediate attestation of students - evaluation of intermediate and final results of training in disciplines (modules) and practical training, results of course design (course work).

Forms, evaluation system, procedure of interim certification of students, as well as the frequency of interim certification of students are established by a local normative act of KNITU-KAI.

The mastering of the presented OP is completed by the state final certification in the form of defense of the final qualification work (GQW), which is mandatory.

6.1 Assessment funds for interim certification and control and measurement materials for current progress control.

The fund of assessment means includes: assessment means for the state final certification; assessment means of interim certification for examinations and credits for disciplines (modules), practices; assessment means of current control (teacher's materials to check the students' mastery of the educational material, including input control; control in practical classes, laboratory work, tasks of educational, industrial practice, etc.). A complete set of assessment tools required to assess the learning outcomes of the discipline (module), practice is stored at the department-developer in paper or electronic form.

6.2 Final State Attestation

State final certification in the direction of 22.03.01 Bachelor's degree includes preparation for the defense of the final qualifying work (GQW) and the defense procedure.

The requirements to the content, scope and structure of the final state certification of graduates (a local act of KNITU-KAI).

The purpose of the GIA is a comprehensive assessment of theoretical knowledge, practical skills and competencies of the graduate obtained during the period of study in accordance with the specifics of this bachelor's program on the example of solving one or more professional tasks.

GEC members in the process of defense on the basis of the report of the student, answers to questions, submitted materials (feedback from the head) can judge the level of training of the student and his readiness for professional activity.

In the report, the learner should:

-briefly characterize the relevance of the topic;

- clearly formulate the purpose and objectives of the work;
- briefly describe what exactly was done in the course of the WRC;
- use in the report all illustrative material submitted for defense;
- clearly formulate the conclusions of the work (with an assessment of the results and the degree of their compliance with the given task).

The results of the defense of the final qualification work are determined by the grades "excellent", "good", "satisfactory", "unsatisfactory" and are announced on the day of the defense of the final qualification work after the protocols of the meetings of the State

Attestation Commission are drawn up in accordance with the established procedure and students' credit books are filled in.

Indicators and criteria for assessing the formation of competencies, assessment scale, typical control questions for assessing the results of mastering the program are given in the GIA FS.

ANNOTATIONS.
to the working programs of disciplines (modules) and practices

B1.B.01 "Philosophy"

1.Purpose of studying the discipline (module)

The main purpose of studying the discipline is the formation of future bachelors philosophical thinking, to correctly navigate in the socio-natural world, to think methodologically competent in mastering academic disciplines and creatively solve scientific, technical and practical problems. The study of philosophy is aimed at developing the skills of critical perception and evaluation of information sources, the ability to logically formulate and argue their own vision of problems and ways to solve them.

2.Objectives of the discipline (module)

The main objectives of the discipline are:

- Improving the philosophical culture of students
- The study of the history and theory of philosophy;
- Familiarization with philosophical, religious and scientific worldviews;
- mastering the basic principles and methods of philosophical cognition;
- introduction to the range of philosophical problems related to the field of future professional activity;
- Gain skills in working with original and adapted philosophical texts.

3.Place of the discipline (module) in the structure of the educational program.

The discipline "Philosophy" is part of the compulsory part of Block 1

4. A graduate who has mastered the discipline shall possess the following competencies:

OK-1 - Possess the ability to use the basics of philosophical knowledge to form a world outlook.
OK-6 - Possess the ability to work in a team, tolerantly accepting social, ethnic, confessional and cultural differences

5.The scope of the academic discipline (module) (indicating the labor intensity of all types of academic work)

Scope of the discipline (module) for full-time education

Types of training work	Total labor intensity (in hours)	
	Total ZE /hours	Semester
		2
Total labor intensity of the discipline (module) in ZE/hour	3ZE/108	108
Contact work of students with the teacher by types of training sessions (classroom work), including:	50,3	50,3
Lectures	34	34
Laboratory work	-	-
Practical exercises	16	16
Coursework (consultation, defense)	-	-
Course project (consultations, defense)	-	-
Pre-exam counseling	-	-
Contact work at the intermediate control	0,3	0,3
Independent work of the student (extracurricular work), including:	57,7	57,7
Working through the training material (self-training)	57,7	57,7
Course project (preparation)	-	-
Coursework (preparation)	-	-
Preparation for interim certification	-	-
Intermediate certification	credit	credit

RPA Developer: E.R. Galimov, Doctor of Technical Sciences, Professor of the Chair of MS&PS, Associate Professor of the Chair of Philosophy, Candidate of Philosophy Sabirzyanov A.M.

B1.B.02 History

1. Purpose of studying the discipline (module)

The main purpose of the discipline "History": to form students' ideas about general regularities and directions of development of world historical processes, to introduce them to the circle of actual historical problems, issues related to the field of future professional activity.
2. Objectives of the discipline (module)
The main objectives of the discipline are: 1) to familiarize with the main stages and key events of world and Russian history; 2) to teach him to identify the interrelation of past and present events, to treat critically the aspects of development and achievements of different civilizations; 3) to form the ability to take a position in discussions, form their own opinion; to cooperate in a group; to possess the skills of individual techniques of studying the problem and decision-making.
3.Place of the discipline (module) in the structure of the educational program of higher education

The discipline "History" is a part of the Basic part of Block 1.
4. A graduate who has mastered the discipline shall possess the following competencies:
OK-2 - ability to analyze the main stages and patterns of historical development of society to form a civic position
OK-6 - ability to work in a team, tolerantly accepting social, ethnic, confessional and cultural differences

5 The scope of the discipline (module) (indicating the labor intensity of all types of academic work).
Scope of the discipline for full-time students

Types of training work	Total labor intensity (in hours)	
	Total ZE /hours	Semester
		2
Total labor intensity of the discipline (module) in ZE/hour	2ZE/72.	72
Contact work of students with the teacher by types of training sessions (classroom work), including:	32,3	32,3
Lectures	16	16
Laboratory work	-	-
Practical exercises	16	16
Coursework (consultation, defense)	-	-
Course project (consultations, defense)	-	-
Pre-exam counseling	-	-
Contact work at the intermediate control	0,3	0,3

Independent work of the student (extracurricular work), including:	39,7	39,7
Working through the training material (self-training)	39,7	39,7
Course project (preparation)	-	-
Coursework (preparation)	-	-
Preparation for interim certification	-	-
Intermediate certification	credit	credit

RPA Developer: E.R. Galimov, Doctor of Technical Sciences, Professor, Department of MS&PB, D.V. Shmelev, Doctor of Science, Professor, ISO Department

B1.B.03 "Foreign Language"

1. Purpose of studying the discipline

- formation of intercultural communicative professionally oriented competence.
- The study of a foreign language is also intended to provide: an increase in the level of learning autonomy, the ability to self-education;
- Development of cognitive and research skills;
- development of information culture, skills of working with foreign-language information on paper and electronic media;
- broadening the outlook and improving the general culture of students;
- fostering tolerance and respect for the spiritual values of different countries and peoples.

2. Objectives of the discipline

- development of foreign language communication skills in all types of speech activity (reading, speaking, listening, writing):
- formation and improvement of listening and pronunciation skills in relation to new language and speech material;
- Formation of skills of all types of reading (studying, browsing, introductory, searching);
- formation of skills of monologic and dialogic statements on the proposed situations; skills of public speaking; skills of drafting simple documents;
- expanding the volume of productive and receptive lexical minimum at the expense of lexical means serving new topics, problems and situations of communication; formation of terminological apparatus

within the framework of the designated topics and problems of communication in the volume of 1200 lexical units.

- correction and development of skills of productive use of basic grammatical forms and constructions;
- formation and improvement of spelling skills in relation to new language and speech material.
- improving the skills of independent work with special (technical) literature in a foreign language in order to obtain professional information and use it in educational and practical activities.

3. place of the discipline in the structure of the educational program of higher education

The discipline "Foreign Language" belongs to the basic part of Block 1 "Disciplines (Modules)" of the Bachelor's degree program and is mandatory for mastering.

4. A graduate who has mastered the discipline shall possess the following competencies:

OK-5-ability to communicate orally and in writing in Russian and foreign languages to solve problems of interpersonal and intercultural interaction

Scope of the discipline (with indication of labor intensity of all types of academic work)

Types of training work	Total labor intensity (in hours)				
	Total ZE /hours	Semester			
		1	2	3	4
Total labor intensity of the discipline (module) in ZE/hour	14/504	3/108	4/144	3/108	4/144
Contact work of students with the teacher by types of training sessions (classroom work), including:		50,3	52,4	50,3	52,4
Lectures					
Laboratory work					
Practical exercises	200	50	50	50	50

Coursework (consultation, defense)					
Course project (consultations, defense)					
Pre-exam counseling	4		2		2
Contact work at the intermediate control	1,4	0,3	0,4	0,3	0,4
Independent work of the student (extracurricular work), including:	298,6	57,7	91,6	57,7	91,6
Working through the training material (self-training)		57,7	58	57,7	58
Course project (preparation)					
Coursework (preparation)					
Preparation for interim certification	67,2		33,6		33,6
Intermediate certification		credit	examination	credit	examination

Developer of RPA: E.Y. Lapteva, Candidate of Pedagogical Sciences, Associate Professor, S.V. Kuryntsev, Candidate of Economic Sciences, Associate Professor.

B1.B.04 "Life Safety"

1.Purpose of studying the discipline (module)

The purpose of training is to form students' fundamental knowledge of the inseparable unity of effective professional activity with the requirements for human safety and security. The realization of these requirements guarantees the preservation of human performance and health, prepares a person to act in extreme conditions.

2. Objectives of the discipline:

- - Study of comfortable (normative) state of habitat in the zones of human labor activity and recreation;
- - Study of the theory of negative habitat impacts of natural, man-made and anthropogenic origin;
- - Knowledge of forecasting the development of negative impacts on humans and the environment, risk assessment and management;
- - To familiarize with the prospects of creation and use of new materials in connection with the most important directions of development of basic industries;
- - Know the measures to protect humans and the environment from negative impacts;
- - Study of modern methods of forming blanks and parts from various materials.
- - To familiarize with the basics of decision-making on the protection of production personnel and population from the possible consequences of accidents, disasters, natural disasters and the use of modern means of defeat, as well as taking measures to eliminate their consequences. Place of the discipline in the structure of the educational program: the discipline "Life Safety" is a part of the basic part.

3. place of the discipline in the structure of the educational program of higher education

The discipline belongs to the basic part of Block 1 "Disciplines (Modules)" of the Bachelor's degree program and is mandatory for mastering.

4. A graduate who has mastered the discipline shall possess the following competencies:

OK-9- readiness to use basic methods of protection of production personnel and population from possible consequences of accidents, catastrophes, natural disasters

OPK-5- ability to apply the principles of rational use of natural resources and environmental protection in practical activities

5. The scope of the discipline (indicating the labor intensity of all types of academic work)

Scope of the discipline for full-time students

	Total labor intensity (in hours)

Types of training work	Total ZE /hours	Semester
		7
Total labor intensity of the discipline (module) in ZE/hour	2 ZE/72.	2 ZE/72.
Contact work of students with the teacher by types of training sessions (classroom work), including:	32,3	32,3
Lectures	16	16
Laboratory work	-	-
Practical exercises	16	16
Coursework (consultation, defense)	-	-
Course project (consultations, defense)	-	-
Pre-exam counseling	-	-
Contact work at the intermediate control	0,3	0,3
Independent work of the student (extracurricular work), including:	39,7	39,7
Working through the training material (self-training)	-	-
Course project (preparation)	-	-
Coursework (preparation)	-	-
Preparation for interim certification	-	-
Intermediate certification		credit

Developer of RPA: V.Yu. Vinogradov, Doctor of Technical Sciences, Associate Professor, Department of MS&PS, I.A. Abrosimov, Candidate of Technical Sciences, Associate Professor, Department of MS&PS, E.R. Galimov, Doctor of Technical Sciences, Professor, Department of MS&PS.

B1.B.05 "Physical Culture and Sports"

1.Purpose of studying the discipline (module)

The purpose of physical education of KNITU-KAI students is the formation of physical culture of personality, preservation and promotion of health, psychophysical preparation for future social and professional activities, inclusion in a healthy lifestyle, systematic physical self-improvement.

2. Objectives of the discipline:

1. to determine the role of physical culture in the development of personality and its preparation for socio-professional activity;

2. to study the scientific and practical foundations of physical education and healthy lifestyle;

3. To form students' motivational and value attitude to physical culture, attitude to a healthy lifestyle, physical self-improvement and self-education, the need for regular physical exercise and sports;

4. Acquire practical skills and abilities ensuring preservation and strengthening of health, mental well-being, development and improvement of psychophysical abilities and properties of personality, self-determination in physical culture;

5. To provide general and professional-applied physical preparedness, which determines the psychophysical readiness of students for their future profession;

6. Gain experience in creative use of physical culture and sports activities to achieve life and professional goals.

3. place of the discipline in the structure of the educational program of higher education

The discipline belongs to the basic part of Block 1 "Disciplines (Modules)" of the Bachelor's degree program and is mandatory for mastering.

4. A graduate who has mastered the discipline shall possess the following competencies:

OK - 8 Ability to use methods and means of physical culture to ensure full-fledged social and professional activity

5. The scope of the discipline (indicating the labor intensity of all types of academic work)

Scope of the discipline for full-time students

Types of training work	Total labor intensity (in hours)		
	Total ZE /hours	Semester	
		1	
Total labor intensity of the discipline (module) in ZE/hour	2 ZE/72.	2 ZE/72.	
Contact work of students with the teacher by types of training sessions (classroom work), including:	16,3	16,3	

Lectures			
Laboratory work	-	-	
Practical exercises	16	16	
Coursework (consultation, defense)	-	-	
Course project (consultations, defense)	-	-	
Pre-exam counseling		-	
Contact work at the intermediate control	0,3	0,3	
Independent work of the student (extracurricular work), including:	55,7	55,7	
Working through the training material (self-training)	55,7	55,7	
Course project (preparation)	-	-	
Coursework (preparation)	-	-	
Preparation for interim certification			
Intermediate certification		credit	

Developer of RPA: Yusupov R.A. Dr. B.Sc., Head of the Department. PhiS, Galimov E.R., Doctor of Technical Sciences, Professor, Department of MS&PB

B1.B.06 "Personal Development"

1.Purpose of studying the discipline (module)

The purpose of studying the discipline is to prepare a specialist, not only well oriented in their future professional activities, but also aimed at personal growth, involving the development of leadership skills, the ability to set and achieve goals, prioritize, work in a team, manage themselves, people and other qualities of a modern competitive personality.

2. Objectives of the discipline:

- Familiarization with the concept of "personality". Study of the components of the inner and outer world of personality, criteria of its maturity.
- Acquisition of knowledge of the basics of the theory and practice of personal development. Mastering the concepts: "personal development", "professional development", "personal growth", "professional growth", "soft-skills skills".
- Identifying the role of "social elevators" in career and personal growth.
- Comprehension of the peculiarities of personality and its mental states, contributing to the acquisition of soft-skills skills, leadership potential, mastering the methods of their diagnostics and self-diagnosis.

- Formation of the ability to take into account personal characteristics of other people, tolerantly perceive their psychological, social, ethnic, professional and cultural differences.
- Familiarization with the content of time management and its principles, formation of motivation for self-organization, mastering effective business communications and time management methods.
- Learning the rules of maintaining physical and mental health of a person as the basis for his/her development.
- Acquisition of an understanding of the basics of the development and application of diagnostic techniques, social-psychological trainings and methods of psychological assistance for personal development.

3. place of the discipline in the structure of the educational program of higher education
The discipline belongs to the basic part of Block 1 "Disciplines (Modules)" of the Bachelor's degree program and is mandatory for mastering.
4. A graduate who has mastered the discipline shall possess the following competencies:
OK-6- ability to work in a team, tolerantly accepting social, ethnic, confessional and cultural differences
OK-7- ability to self-organization and self-education

5. The scope of the discipline (indicating the labor intensity of all types of academic work)

Types of training work	Total labor intensity (in hours)	
	Total hours (ZE)	Semester
		1
Total labor intensity of the discipline (module) in ZE/hour	2 ZE/72.	2 ZE/72.
Contact work of students with the teacher by types of training sessions (classroom work), including:	32,3	32,3
Lectures	16	16
Laboratory work	-	-
Practical exercises	16	16

Coursework (consultation, defense)	-	-
Course project (consultations, defense)	-	-
Pre-exam counseling	-	-
Contact work at the intermediate control	0,3	0,3
Independent work of the student (extracurricular work), including:	39,7	39,7
Working through the training material (self-training)	39,7	39,7
Course project (preparation)	-	-
Coursework (preparation)	-	-
Preparation for interim certification	-	-
Intermediate certification		credit

RPA developer: A.R. Karimova, Ph.D. in Psychology, Associate Professor, Chair of ET&SD, E.R. Galimov, Ph.

B1.B.07 "Higher Mathematics"

1.Purpose of studying the discipline (module)

The main purpose of studying the discipline "Higher Mathematics" is the formation of mathematical culture in future bachelors, including a clear understanding of the necessity of the mathematical component in the general training of a bachelor, the development of ideas about the role and place of mathematics in modern civilization and in world culture, the ability to think logically, operate with abstract objects and be correct in the use of mathematical concepts and symbols to express quantitative and qualitative relationships.

2. Objectives of the discipline:

- Formation of students' basic knowledge of the following sections: linear algebra (including linear mappings), vector algebra, analytical geometry (including curves and surfaces of the second order), mathematical analysis (including differential geometry), elements of functional analysis (elements of topologies), probability theory and mathematical statistics, discrete mathematics (logical calculus, graphs, elements of combinatorics).
- Formation of skills of using methods of linear algebra, vector algebra, analytical geometry, mathematical analysis, elements of functional analysis, methods of probability theory and mathematical statistics,

discrete mathematics in technical applications.

- Developing the ability to use mathematical concepts and symbols correctly
- Formation of the ability to use modern information and educational technologies in independent work

3. place of the discipline in the structure of the educational program of higher education
The discipline belongs to the basic part of Block 1 "Disciplines (Modules)" of the Bachelor's degree program and is mandatory for mastering.
4. A graduate who has mastered the discipline shall possess the following competencies:
OK-6- ability to work in a team, tolerantly accepting social, ethnic, confessional and cultural differences
OK-7- ability to self-organization and self-education

5. The scope of the discipline (indicating the labor intensity of all types of academic work)

Types of training work	**Total labor intensity (in hours)**			
	Total ZE /hours	**Semester**		
		1	**2**	**3**
Total labor intensity of the discipline (module) in ZE/hour	**18/648**	**7/252**	**7/252**	**7/252**
Contact work of students with the teacher by types of training sessions (classroom work), including:	***356***	***136***	***136***	***84***
Lectures	170	68	68	34
Laboratory work				
Practical exercises	186	68	68	50
Coursework (consultation, defense)				
Course project (consultations, defense)				
Pre-exam counseling	4	2	2	
Contact work at the intermediate control	1,7	0,7	0,7	0,3

Independent work of the student (extracurricular work), including:	209	57,7	91,6	59,7
Working through the training material (self-training)	286,2	113,3	113,3	59,7
Course project (preparation)				
Coursework (preparation)				
Preparation for interim certification	67,2	33,6	33,6	
Intermediate certification		Pass, exam	Pass, exam	credi t

Developer of the RPA: Sidorov I.N., Doctor of Technical Sciences, Professor, Department of TM&VM, Takseitov R.R., Candidate of Technical Sciences, Associate Professor,

B1.B.08 "Physics"

1.Purpose of studying the discipline (module)
The purpose of studying the discipline is to form in future bachelors fundamental knowledge of physics, necessary for the study of subsequent professional disciplines and in future professional activities. 2.

2. Objectives of the discipline:
- Study of basic physical phenomena; mastering fundamental concepts, laws, theories of classical and modern physics;
- Formation of scientific worldview and modern physical thinking;
- Mastering the techniques and methods of solving specific problems from different areas of physics;
- Familiarization with modern scientific equipment, methods of physical research, formation of skills of physical experimentation and preparation of scientific and technical reports;
- Formation of the graduate's ability to use the basic laws of physics in professional activities, application of theoretical and experimental research methods, participation in the development of mathematical and physical models of processes.

3. place of the discipline in the structure of the educational program of higher education
The discipline belongs to the basic part of Block 1 "Disciplines (Modules)" of the Bachelor's degree program and is mandatory for mastering.
4. A graduate who has mastered the discipline shall possess the following competencies:
OPK-2: ability to use in professional activity knowledge of approaches and methods of obtaining results in theoretical and experimental research
OPK-3:readiness to apply fundamental mathematical, natural science and general engineering knowledge in professional activity
OPK-4: ability to combine theory and practice to solve engineering problems

5. The scope of the discipline (indicating the labor intensity of all types of academic work)

Types of training work	Total labor intensity (in hours)			
	Total ZE /hours	Semester		
		1	2	3
Total labor intensity of the discipline (module) in ZE/hour	14ZE/50 4	5 ZE/180 .	5 ZE/180 .	4 ZE/14 4.
Contact work of students with the teacher by types of training sessions (classroom work), including:	*197,1*	*68,4*	*68,4*	*68,3*
Lectures	102	34	34	34
Laboratory work	48	16	16	16
Practical exercises	48	16	16	16
Coursework (consultation, defense)	-	-	-	-
Course project (consultations, defense)	-	-	-	-
Pre-exam counseling	4	2	2	
Contact work at the intermediate control	1,1	0,4	0,4	0,3

Independent work of the student (extracurricular work), including:	*239,7*	*111,6*	*111,6*	*77,7*
Working through the training material (self-training)	133,3	78	78	77,7
Course project (preparation)	-	-	-	-
Coursework (preparation)	-	-	-	-
Preparation for interim certification	67,2	33,6	33,6	
Intermediate certification		examination	examination	credit

RPA Developer: R.H. Makayeva, Doctor of Technical Sciences, Professor, I.N. Sidorov, Doctor of Technical Sciences, Professor, Department of TM&VM.

B1.B.09 Metrology, Standardization and Certification

1.Purpose of studying the discipline (module)

The main purpose of studying the discipline is to master the basic provisions of metrology and metrological support, formation of ideas about modern methods and means in the field of metrology, accuracy, standardization and certification.

2. Objectives of the discipline:

1. study of basic provisions of metrology, principles and methods of processing and presentation of measurement results;
2. the ability to apply a unified system of norming and standardization of measurement accuracy indicators, as well as the possession of methods for selecting measuring instruments in testing and control of products;
3. ability to verify compliance of the values of measured and controlled parameters of products and technological processes specified in the technical documentation with the service purpose of the part and relevant national standards;
4. development of a systematic approach to solving metrological problems in the field of organization and implementation of quality control of products, materials, components, production control of technological processes.

3. place of the discipline in the structure of the educational program of higher education

The discipline belongs to the basic part of Block 1 "Disciplines (Modules)" of the Bachelor's degree program and is mandatory for mastering.

4. A graduate who has mastered the discipline shall possess the following competencies:

OPK-2: ability to use in professional activity knowledge of approaches and methods of obtaining results in theoretical and experimental research

5. The scope of the discipline (indicating the labor intensity of all types of academic work)

Types of training work	Total labor intensity (in hours)	
	Total ZE /hours	Semester
		3
Total labor intensity of the discipline (module) in ZE/hour	3ZE/108	108
Contact work of students with the teacher by types of training sessions (classroom work), including:	32,3	32,3
Lectures	16	16
Laboratory work	-	-
Practical exercises	16	16
Coursework (consultation, defense)	-	-
Course project (consultations, defense)	-	-
Pre-exam counseling	-	-
Contact work at the intermediate control	0,3	0,3
Independent work of the student (extracurricular work), including:	75,7	75,7
Working through the training material (self-training)	75,7	75,7
Course project (preparation)	-	-
Coursework (preparation)	-	-
Preparation for interim certification	-	-
Intermediate certification	credit	credit

Developer of RPA: Sidorov I.N., Doctor of Technical Sciences, Professor of the Department of TM&VM, Senior Lecturer D.A. Bulashov. Bulashov D.A.

B1.B.10.01 "Theory of research problem solving"

1.Purpose of studying the discipline (module)

The main purpose of studying the discipline is to form in future bachelors actual practical competencies that allow them to successfully solve inventive problems related to research and calculation-analytical, production and design-technological preparation of development, creation and production of products from new materials.2. 2. Tasks of the discipline:

assimilation of knowledge about the basic principles and laws of development of technical systems;

- mastering the methods of inventive competence development, including those based on TRIZ technology;
- developing teamwork skills.

3. place of the discipline in the structure of the educational program of higher education

The discipline belongs to the basic part of Block 1 "Disciplines (Modules)" of the Bachelor's degree program and is mandatory for mastering.

4. A graduate who has mastered the discipline shall possess the following competencies:

OK-7 ability to self-organization and self-education

OPK-2 - ability to use in professional activity knowledge of approaches and methods of obtaining results in theoretical and experimental research

OPK-4 - ability to combine theory and practice to solve engineering problems

5. The scope of the discipline (indicating the labor intensity of all types of academic work)

Types of training work	**Total labor intensity (in hours)**		
	Total ZE /hours	**Semester**	
		2	
Total labor intensity of the discipline (module) in ZE/hour	2 ZE/72.	2 ZE/72.	
Contact work of students with the teacher by types of training sessions (classroom work), including:	***16,3***	***16,3***	
Lectures	16	16	

Laboratory work	-	-	
Practical exercises	-	-	
Coursework (consultation, defense)	-	-	
Course project (consultations, defense)	-	-	
Pre-exam counseling	-	-	
Contact work at the intermediate control	0,3	0,3	
Independent work of the student (extracurricular work), including:	55,7	55,7	
Working through the training material (self-training)	55,7	55,7	
Course project (preparation)	-	-	
Coursework (preparation)	-	-	
Preparation for interim certification	-	-	
Intermediate certification		credit	

RPA Developer: V.A. Morozov, Associate Professor of the Department of Chemical Technology and New Materials, Lomonosov Moscow State University, Lopatin. M.V. Lomonosov Moscow State University, Lopatin. A.A. Lopatin, Candidate of Technical Sciences, Associate Professor of the Department of Chemical Technology and New Materials

B1.B.10.02 "Basics of project activity"
1.Purpose of studying the discipline (module)
As a result of mastering this discipline, the graduate acquires knowledge, skills and abilities that ensure the achievement of the following objectives: Familiarization of graduates with the tools for organizing project and management activities in the implementation of innovative and other commercial projects in the direction of training.
2. Objectives of the discipline:
assimilation by students of knowledge about the basic principles and laws of development of technical systems;
- mastering the methods of developing inventive competence;
- developing teamwork skills.

3. place of the discipline in the structure of the educational program of higher education
The discipline belongs to the basic part of Block 1 "Disciplines (Modules)" of the Bachelor's degree program and is mandatory for mastering.
4. A graduate who has mastered the discipline shall possess the following competencies:
OPK-2 - ability to use in professional activity knowledge of approaches and methods of obtaining results in theoretical and experimental research
OPK-4 - ability to combine theory and practice to solve engineering problems

5. The scope of the discipline (indicating the labor intensity of all types of academic work)

Types of training work	**Total labor intensity (in hours)**		
	Total ZE /hours	**Semester**	
		5	
Total labor intensity of the discipline (module) in ZE/hour	2 ZE/72.	2 ZE/72.	
Contact work of students with the teacher by types of training sessions (classroom work), including:	***32,3***	***32,3***	
Lectures	32	32	
Laboratory work	-	-	
Practical exercises	-	-	
Coursework (consultation, defense)	-	-	
Course project (consultations, defense)	-	-	
Pre-exam counseling	-	-	
Contact work at the intermediate control	0,3	0,3	
Independent work of the student (extracurricular work), including:	39,7	39,7	
Working through the training material (self-training)	39,7	39,7	
Course project (preparation)	-	-	
Coursework (preparation)	-	-	

Preparation for interim certification	-	-	
Intermediate certification		credit	

Developer of RPA: V.A. Morozov, Associate Professor of the Department of Chemical Technology and New Materials, Lomonosov Moscow State University. M.V. Lomonosov Moscow State University

B1.B.10.03 "Enterprise Economics and Digital Production"

1.Purpose of studying the discipline (module)
As a result of mastering this discipline, the graduate acquires knowledge, skills and abilities that ensure the achievement of the following objectives: Familiarization of graduates with the tools for organizing project and management activities in the implementation of innovative and other commercial projects in the direction of training.
2. Objectives of the discipline:
The main objectives of the discipline are:
Obtain economic knowledge and use it in practical activities.
3. place of the discipline in the structure of the educational program of higher education
The discipline belongs to the basic part of Block 1 "Disciplines (Modules)" of the Bachelor's degree program and is mandatory for mastering.
4. A graduate who has mastered the discipline shall possess the following competencies:
OPK-2 - ability to use in professional activity knowledge of approaches and methods of obtaining results in theoretical and experimental research
OPK-4 - ability to combine theory and practice to solve engineering problems

5. The scope of the discipline (indicating the labor intensity of all types of academic work)

Types of training work	**Total labor intensity (in hours)**		
	Total ZE /hours	**Semester**	
		4	
Total labor intensity of the discipline (module) in ZE/hour	2 ZE/72.	2 ZE/72.	

Contact work of students with the teacher by types of training sessions (classroom work), including:	***32,3***	***32,3***	
Lectures	32	32	
Laboratory work	-	-	
Practical exercises	-	-	
Coursework (consultation, defense)	-	-	
Course project (consultations, defense)	-	-	
Pre-exam counseling	-	-	
Contact work at the intermediate control	0,3	0,3	
Independent work of the student (extracurricular work), including:	39,7	39,7	
Working through the training material (self-training)	39,7	39,7	
Course project (preparation)	-	-	
Coursework (preparation)	-	-	
Preparation for interim certification	-	-	
Intermediate certification		credit	

Developer of the RPA: E.R. Galimov, Prof. of the Department of MS&PSS. Department of MS&PB

B1.B.11.01 "Descriptive Geometry and Engineering Graphics"
1.Purpose of studying the discipline (module)
The purpose of studying the educational discipline of NG and IG is the formation of basic knowledge for mastering special disciplines and the formation of professional competencies.
2. Objectives of the discipline:
The main objectives of the discipline are:
- mastering the theoretical foundations of drawing construction.
- mastering the basics of development of design documentation for various purposes in compliance with the requirements of the ESCD standards.

3. place of the discipline in the structure of the educational program of higher education
The discipline belongs to the basic part of Block 1 "Disciplines (Modules)" of the Bachelor's degree program and is mandatory for mastering.
4. A graduate who has mastered the discipline shall possess the following competencies:
OPK-1- ability to solve standard tasks of professional activity on the basis of information and bibliographic culture with the use of information and communication technologies and taking into account the basic requirements of information security
OPK-4 - ability to combine theory and practice to solve engineering problems

5. The scope of the discipline (indicating the labor intensity of all types of academic work)

Types of training work	**Total labor intensity (in hours)**	
	Total ZE /hours	**Semester 1**
Total labor intensity of the discipline (module) in ZE/hour	4 ZE/144.	4 ZE/144.
Contact work of students with the teacher by types of training sessions (classroom work), including:	***70,4***	***70,4***
Lectures	34	34
Laboratory work	32	32
Practical exercises	-	-
-	-	-
Course project (consultations, defense)	-	-
Pre-exam counseling	2	2
Contact work at the intermediate control	0.4	0.4
Independent work of the student (extracurricular work), including:	*75,6*	*75,6*

Working through the training material (self-training)	42	42
Course project (preparation)	-	-
Coursework (preparation)	-	-
Preparation for interim certification	33,6	33,6
Intermediate certification		**examination**

RPA Developer: N.Y. Galimova, Candidate of Technical Sciences, Associate Professor, Department of M and IG.

B1.B.11.02 "Computer Graphics"

1.Purpose of studying the discipline (module)

The purpose of studying the academic discipline "Computer Graphics" is the formation of basic knowledge for mastering special disciplines and the formation of professional competencies.

2. Objectives of the discipline:

The main objectives of the discipline are:

- familiarization of students with ways of automation of engineering activity, processing of geometric information, development of skills of drawing execution on PC.
- mastering the requirements for sketch technical documentation and working drawings of parts.

3. place of the discipline in the structure of the educational program of higher education

The discipline belongs to the basic part of Block 1 "Disciplines (Modules)" of the Bachelor's degree program and is mandatory for mastering.

4. A graduate who has mastered the discipline shall possess the following competencies:

OPK-1- ability to solve standard tasks of professional activity on the basis of information and bibliographic culture with the use of information and communication technologies and taking into account the basic requirements of information security
OPK-4 - ability to combine theory and practice to solve engineering problems

5. The scope of the discipline (indicating the labor intensity of all types of academic
work)

Types of training work	**Total labor intensity (in hours)**	
	Total ZE /hours	**Semester**
		2
Total labor intensity of the discipline (module) in ZE/hour	2 ZE/72.	2ZE/72.
Contact work of students with the teacher by types of training sessions (classroom work), including:	***32,3***	***32,3***
Lectures	-	-
Laboratory work	32	32
Practical exercises	-	-
-	-	-
Course project (consultations, defense)	-	-
Pre-exam counseling	-	-
Contact work at the intermediate control	0.3	0.3
Independent work of the student (extracurricular work), including:	***39,7***	***39,7***
Working through the training material (self-training)	39.7	39.7
Course project (preparation)	-	-
Coursework (preparation)	-	-
Preparation for interim certification	-	-

Intermediate certification		credit

RPA Developer: N.Y. Galimova, Candidate of Technical Sciences, Associate Professor, Department of M and IG.

B1.B.12.01 "Informatics"

1.Purpose of studying the discipline (module)

The aim of training is to form students' fundamental knowledge about the nature and properties of materials, about the dependencies of their properties on composition and structure, about the regularities of transformations in metals and alloys in various thermophysical conditions and processes occurring in materials under load to form skills of scientifically justified choice of materials, application of highly effective methods of their processing and purposeful use in structures with a high degree of reliability and durability.

2. Objectives of the discipline:

The main objectives of the discipline are:

- Study of the essence and importance of information in the development of modern information society;
- Acquisition of skills to work with information in computer networks;
- acquiring skills of competent and rational use of a computer as an information management tool;
- Acquisition of skills of using modern information and communication technologies and global resources in research and calculation-analytical activities in the field of materials science and technology of materials;
- Acquisition of skills of data collection, analysis of scientific and technical information, development and use of technical documentation, basic normative documents on intellectual property issues, preparation of documents for patenting, registration of know-how.

3. place of the discipline in the structure of the educational program of higher education

The discipline belongs to the basic part of Block 1 "Disciplines (Modules)" of the Bachelor's degree program and is mandatory for mastering.

4. A graduate who has mastered the discipline shall possess the following competencies:

OPK-1- ability to solve standard tasks of professional activity on the basis of information and bibliographic culture with the use of information and communication technologies and taking into account the basic requirements of information security

5. The scope of the discipline (indicating the labor intensity of all types of academic
work)

Types of training work	**Total labor intensity (in hours)**	
	Total ZE /hours	**Semester**
		1
Total labor intensity of the discipline (module) in ZE/hour	2 ZE/72.	2 ZE/72.
Contact work of students with the teacher by types of training sessions (classroom work), including:	***32,3***	***32,3***
Lectures	-	-
Laboratory work	32	32
Practical exercises	-	-
Coursework (consultation, defense)	-	-
Course project (consultations, defense)	-	-
Pre-exam counseling	-	-
Contact work at the intermediate control	0,3	0,3
Independent work of the student (extracurricular work), including:	***39,7***	***39,7***
Working through the training material (self-training)	20	20
Course project (preparation)	-	-
Coursework (preparation)	-	-
Preparation for interim certification	19,7	19,7
Intermediate certification		credit

Developer of RPA: A.V. Belyaev, Candidate of Technical Sciences, Associate Professor, Candidate of Technical Sciences, Associate Professor, Department of MS&PB, E.S. Mukhametshina, Candidate of Technical Sciences, Associate Professor, Department of MS&PB.

B1.B.12.01 "Informatics"

1.Purpose of studying the discipline (module)

The aim of training is to form students' fundamental knowledge about the nature and properties of materials, about the dependencies of their properties on composition and structure, about the regularities of transformations in metals and alloys in various thermophysical conditions and processes occurring in materials under load to form skills of scientifically justified choice of materials, application of highly effective methods of their processing and purposeful use in structures with a high degree of reliability and durability.

2. Objectives of the discipline:

The main objectives of the discipline are:

- Study of the essence and importance of information in the development of modern information society;

- Acquisition of skills to work with information in computer networks;

- acquiring skills of competent and rational use of a computer as an information management tool;

- Acquisition of skills of using modern information and communication technologies and global resources in research and calculation-analytical activities in the field of materials science and technology of materials;

- Acquisition of skills of data collection, analysis of scientific and technical information, development and use of technical documentation, basic normative documents on intellectual property issues, preparation of documents for patenting, registration of know-how.

3. place of the discipline in the structure of the educational program of higher education

The discipline belongs to the basic part of Block 1 "Disciplines (Modules)" of the Bachelor's degree program and is mandatory for mastering.

4. A graduate who has mastered the discipline shall possess the following competencies:

OPK-1- ability to solve standard tasks of professional activity on the basis of information and bibliographic culture with the use of information and communication technologies and taking into account the basic requirements of information security

5. The scope of the discipline (indicating the labor intensity of all types of academic
work)

Types of training work	**Total labor intensity (in hours)**	
	Total ZE /hours	**Semester**
		1
Total labor intensity of the discipline (module) in ZE/hour	2 ZE/72.	2 ZE/72.
Contact work of students with the teacher by types of training sessions (classroom work), including:	***32,3***	***32,3***
Lectures	-	-
Laboratory work	32	32
Practical exercises	-	-
Coursework (consultation, defense)	-	-
Course project (consultations, defense)	-	-
Pre-exam counseling	-	-
Contact work at the intermediate control	0,3	0,3
Independent work of the student (extracurricular work), including:	***39,7***	***39,7***
Working through the training material (self-training)	20	20
Course project (preparation)	-	-
Coursework (preparation)	-	-
Preparation for interim certification	19,7	19,7
Intermediate certification		credit

RPA Developer: A.V. Belyaev, Candidate of Technical Sciences, Associate Professor, Candidate of Technical Sciences, Associate Professor, Department of MS&PS, E.S. Mukhametshina, Candidate of Technical Sciences, Associate Professor, Department of MS&PS

B1.B.12.02 "Computer-aided design systems"

1.Purpose of studying the discipline (module)

The purpose of training is to form students' knowledge of computer-aided design systems (CAD), types of CAD support, process modeling and optimization, levels of automation, CAD software, process design issues, prototyping.

2. Objectives of the discipline:

The main objectives of the discipline are:

– Study of basic CAD terms and definitions.

– Study of types of CAD support (mathematical, software, technical, information, etc.).

– Familiarization with the concepts and essence of modeling, finite element and difference methods.

– Study of basic software in the field of CAD.

– Acquisition of skills in 3D modeling systems.

3. place of the discipline in the structure of the educational program of higher education

The discipline belongs to the basic part of Block 1 "Disciplines (Modules)" of the Bachelor's degree program and is mandatory for mastering.

4. A graduate who has mastered the discipline shall possess the following competencies:

OPK-1- ability to solve standard tasks of professional activity on the basis of information and bibliographic culture with the use of information and communication technologies and taking into account the basic requirements of information security

5. The scope of the discipline (indicating the labor intensity of all types of academic
work)

Types of training work	**Total labor intensity (in hours)**	
	Total ZE /hours	**Semester**
		4
Total labor intensity of the discipline (module) in ZE/hour	2 ZE/72.	2 ZE/72.
Contact work of students with the teacher by types of training sessions (classroom work), including:	***32,3***	***32,3***
Lectures	–	–
Laboratory work	32	32
Practical exercises	–	–
Coursework (consultation, defense)	–	–
Course project (consultations, defense)	–	–
Pre-exam counseling	–	–
Contact work at the intermediate control	0,3	0,3
Independent work of the student (extracurricular work), including:	***39,7***	***39,7***
Working through the training material (self-training)	39,7	39,7
Course project (preparation)	–	–
Coursework (preparation)	–	–
Preparation for interim certification	–	–
Intermediate certification	credit	credit

RPA Developer: A.V. Belyaev, Candidate of Technical Sciences, Associate Professor,

B1.B.12.03 "Application software packages in professional activities"

1.Purpose of studying the discipline (module)

The main purpose of the discipline is to familiarize students with the basic provisions of informatics, modern computer technologies; formation of skills to apply these technologies in scientific research, solving problems of obtaining materials with high performance properties, improvement, development of new high-performance technologies.

.

2. Objectives of the discipline:

mastering of general provisions of informatics, computer and information technologies, their application in scientific work and in practice;

formation of skills of using modern computer and information technologies in the performance of research and development, solving production problems;

consolidation of these skills by performing tasks at practical classes and laboratory works.

3. place of the discipline in the structure of the educational program of higher education

The discipline belongs to the basic part of Block 1 "Disciplines (Modules)" of the Bachelor's degree program and is mandatory for mastering.

4. A graduate who has mastered the discipline shall possess the following competencies:

OPK-2-ability to use in professional activity knowledge of approaches and methods of obtaining results in theoretical and experimental research

5. The scope of the discipline (indicating the labor intensity of all types of academic
work)

Types of training work	**Total labor intensity (in hours)**		
	Total ZE /hours	**Semester**	**Semester**
		5	**6**

Total labor intensity of the discipline (module) in ZE/hour	4 ZE/144.	2 ZE/72.	2 ZE/72.
Contact work of students with the teacher by types of training sessions (classroom work), including:	***64,6***	***32,3***	***32,3***
Lectures	–	–	–
Laboratory work	64	32	32
Practical exercises	–	–	–
Coursework (consultation, defense)	–	–	–
Course project (consultations, defense)	–	–	–
Pre-exam counseling	–	–	–
Contact work at the intermediate control	0,6	0,3	0,3
Independent work of the student (extracurricular work), including:	***79,4***	***39,7***	***39,7***
Working through the training material (self-training)	79,4	39,7	39,7
Course project (preparation)	–	–	–
Coursework (preparation)	–	–	–
Preparation for interim certification	–	–	–
Intermediate certification		credit	credit

RPA Developer: A.V. Belyaev, Candidate of Technical Sciences, Associate Professor, Department of MS&PS, V.L. Fedyaev, Doctor of Technical Sciences, Professor, Department of MS&PS.

B1.B.13 "Chemistry"

1.Purpose of studying the discipline (module)

formation of students' understanding of the theoretical foundations of chemistry as a system of sciences for the subsequent use of this knowledge in the study of other disciplines, for professional competence and ensuring human safety in the modern world.

2. Objectives of the discipline:

study of basic concepts, laws and models of chemical systems,

- the study of the reactivity of substances.

3. place of the discipline in the structure of the educational program of higher education

The discipline belongs to the basic part of Block 1 "Disciplines (Modules)" of the Bachelor's degree program and is mandatory for mastering.

4. A graduate who has mastered the discipline shall possess the following competencies:

OPK-2-ability to use in professional activity knowledge of approaches and methods of obtaining results in theoretical and experimental research

OPK-3- readiness to apply fundamental mathematical, natural-scientific and general engineering knowledge in professional activities

5. The scope of the discipline (indicating the labor intensity of all types of academic work)

Types of training work	**Total labor intensity (in hours)**	
	Total ZE /hours	**Semester**
		2
Total labor intensity of the discipline (module) in ZE/hour	**3ZE/108.**	**108**
Contact work of students with the teacher by types of training sessions (classroom work), including:	48,3	48,3
Lectures	16	16
Laboratory work	16	16
Practical exercises	16	16
Coursework (consultation, defense)	-	-
Course project (consultations, defense)	-	-
Pre-exam counseling	-	-
Contact work at the intermediate control	0,3	0,3
Independent work of the student (extracurricular work), including:	59,7	59,7
Working through the training material (self-training)	59,7	59,7
Course project (preparation)	-	-
Coursework (preparation)	-	-
Preparation for interim certification	-	-
Intermediate certification	**credit**	**credit**

R.S. Davletbaev, Doctor of Technical Sciences, Associate Professor, Department of MS&PB. Grigorieva I.G., Associate Professor, Department of Chemical Engineering.

B1.B.14 "Theoretical Mechanics"

1.Purpose of studying the discipline (module)

The main purpose of the study of theoretical mechanics (TM) is the formation of future bachelors knowledge of the basic laws of mechanics, the ability to solve problems of statics, kinematics and dynamics, the ability to select adequate mechanical models of designed technical systems, the ability to use the laws and methods of TM in the study of other disciplines and for professional competence.

2. Objectives of the discipline:

The main objectives of the discipline are:

- Learning the basic concepts of statics, equations of equilibrium, and how to use them to determine bond reactions;
- study of basic concepts of kinematics, methods of setting motion and determining kinematic parameters of motion of a material point and a solid body;
- study of axioms of dynamics of a material point (Newton's laws), general theorems of dynamics of a material system and their use for solving applied problems and building mathematical models of motion of real mechanical objects.

3. place of the discipline in the structure of the educational program of higher education

The discipline belongs to the basic part of Block 1 "Disciplines (Modules)" of the Bachelor's degree program and is mandatory for mastering.

4. A graduate who has mastered the discipline shall possess the following competencies:

OPK-2-ability to use in professional activity knowledge of approaches and methods of obtaining results in theoretical and experimental research

5. The scope of the discipline (indicating the labor intensity of all types of academic
work)

Types of training work	**Total labor intensity (in hours)**	
		Semester

	Total ZE /hours	2	3
Total labor intensity of the discipline (module) in ZE/hour	**8 ZE/288.**	**4 ZE/144.**	**4 ZE/144.**
Contact work of students with the teacher by types of training sessions (classroom work), including:	**122,8**	**70,4**	**52,4**
Lectures	**68**	**34**	**34**
Laboratory work	-	-	-
Practical exercises	**50**	**34**	**16**
Coursework (consultation, defense)	-	-	-
Course project (consultations, defense)	-	-	-
Pre-exam counseling	**4**	**2**	**2**
Contact work at the intermediate control	**0,8**	**0,4**	**0,4**
Independent work of the student (extracurricular work), including:	**165,2**	**73,6**	**91,6**
Working through the training material (self-training)	**98**	**40**	**58**
Course project (preparation)	-	-	-
Coursework (preparation)	-	-	-
Preparation for interim certification	**67,2**	**33,6**	**33,6**
Intermediate certification		**examination**	**examination**

Developer of RPA: Sidorov I.N., Doctor of Physical and Mathematical Sciences, Professor, Department of TIPM&M,

B1.B.15 "Theory of mechanisms and machines"

1.Purpose of studying the discipline (module)

The purpose of studying the discipline is to form at future bachelors the basic and most important ideas about modern methods in the field of design of typical machine assemblies and elements of aircraft structures

2. Objectives of the discipline:

1. Collection and analysis of initial information for the development of product designs (parts, units, assemblies) of aircraft and their systems;

2. Mastery of modern methods of structural, kinematic and dynamic synthesis and analysis of schemes of various aircraft mechanisms;

3. Design of products and systems of aircraft equipment in accordance with the technical assignment using information technologies and means of automation of design work;
4. Ability to design elements of machines and aircraft structures taking into account strength, stability and durability.
3. place of the discipline in the structure of the educational program of higher education
The discipline belongs to the basic part of Block 1 "Disciplines (Modules)" of the Bachelor's degree program and is mandatory for mastering.
4. A graduate who has mastered the discipline shall possess the following competencies:
OPK-3-readiness to apply fundamental mathematical, natural science and general engineering knowledge in professional activity

5. The scope of the discipline (indicating the labor intensity of all types of academic
work)

Types of training work	**Total labor intensity (in hours)**	
	Total /hours ZE	**Semester 4**
Total labor intensity of the discipline (module) in ZE/hour	**4ZE/144.**	**144**
Contact work of students with the teacher by types of training sessions (classroom work), including:	68.6	68,6
Lectures	34	34
Laboratory work	16	16
Practical exercises	16	16
Coursework (consultation, defense)	2,3	2,3
Course project (consultations, defense)	-	-
Pre-exam counseling	-	-
Contact work at the intermediate control	0,3	0,3
Independent work of the student (extracurricular work), including:	**75,4**	**75.4**
Working through the training material (self-training)	41,7	41,7
Course project (preparation)	-	-
Coursework (preparation)	33,7	33,7

Preparation for interim certification	-	-
Intermediate certification	**credit**	**credit**

Developer of RPA: Galimova N.Ya. Candidate of Technical Sciences, Associate Professor, Department of M and IG,

B1.B.16 "Machine parts"

1.Purpose of studying the discipline (module)

The purpose of studying the discipline "Machine parts" is to form the future bachelors' understanding of modern methods in the field of design of typical assemblies and elements of machines and mechanisms.

2. Objectives of the discipline:

The main objectives of the discipline are

- collection and analysis of initial information for the development of designs of products (parts, units, assemblies) of general mechanical engineering;
- mastering of modern methods of structural, kinematic and dynamic synthesis and analysis of schemes of various machine mechanisms;
- knowledge of the basic principles of designing products and systems of obo-equipment of machines and mechanisms of general engineering in accordance with the technical assignment with the use of information technologies and means of automation of design works;
- ability to maintain technological equipment in the implementation of production processes.

3. place of the discipline in the structure of the educational program of higher education

The discipline belongs to the basic part of Block 1 "Disciplines (Modules)" of the Bachelor's degree program.

4. A graduate who has mastered the discipline shall possess the following competencies:

OPK-2- ability to use in professional activity knowledge of approaches and methods of obtaining results in theoretical and experimental research

OPK-4 ability to combine theory and practice to solve engineering problems

5. The scope of the discipline (indicating the labor intensity of all types of academic
work)

	Total labor intensity (in hours)

Types of training work	**Total ZE /hours**	**Semester**	
		7	**8**
Total labor intensity of the discipline (module) in ZE/hour	8 ZE/288.	5ZE/180.	3 ZE/108.
Contact work of students with the teacher by types of training sessions (classroom work), including:	***87,7***	***68,4***	***19,3***
Lectures	34	34	-
Laboratory work	32	32	-
Practical exercises	16	-	16
Coursework (consultation, defense)			-
Course project (consultations, defense)	3,3		3,3
Pre-exam counseling	2	2	
Contact work at the intermediate control	0,4	0,4	
Independent work of the student (extracurricular work), including:	***200,3***	***111,6***	***88,7***
Working through the training material (self-training)	134	78	56
Course project (preparation)	32,7	-	32,7
Coursework (preparation)		-	-
Preparation for interim certification	33,6	33,6	
Intermediate certification		examination	Pass with a grade

Developer of RPA: Galimova N.Ya. Candidate of Technical Sciences, Associate Professor, Department of M and IG,

B1.B.17 "Materials in Engineering"

1.Purpose of studying the discipline (module)
The purpose of studying the discipline is to form the future bachelors' knowledge about the areas and spheres of application of metal materials

used for the manufacture of structures, products, assemblies, parts and units of machine-building, aircraft, shipbuilding purposes, methods of their production and processing; abilities to combine theory and practice to solve engineering problems related to the choice of materials, depending on the field of their application and purpose.

. Discipline Objectives:
The main objectives of the discipline are:
- study of the structure, properties and applications of alloys based on iron, titanium, aluminum, magnesium, nickel;
- study of the influence of alloying elements on operational and technological properties of alloys;
- The ability to select the material of a structure or part depending on the required performance characteristics;
- possession of skills of choosing the optimal technology of processing metals and alloys.

3. place of the discipline in the structure of the educational program of higher education
The discipline belongs to the basic part of Block 1 "Disciplines (Modules)" of the Bachelor's degree program.

4. A graduate who has mastered the discipline shall possess the following competencies:
OPK-4 ability to combine theory and practice to solve engineering problems

5. The scope of the discipline (indicating the labor intensity of all types of academic
work)

Types of training work	**Total labor intensity (in hours)**	
	Total ZE /hours	**Semester**
		5
Total labor intensity of the discipline (module) in ZE/hour	2 ZE/72.	2 ZE/72.
Contact work of students with the teacher by types of training sessions (classroom work), including:	32,3	32,3

Lectures	16	16
Laboratory work	-	-
Practical exercises	16	16
Coursework (consultation, defense)	-	-
Course project (consultations, defense)	-	-
Pre-exam counseling	-	-
Contact work at the intermediate control	0,3	0,3
Independent work of the student (extracurricular work), including:	39,7	39,7
Working through the training material (self-training)	39,7	39,7
Course project (preparation)	-	-
Coursework (preparation)	-	-
Preparation for interim certification	-	-
Intermediate certification		credit

Developer of the RPA: S. Kuryntsev. V., Candidate of Economic Sciences, Associate Professor, Department of MS&PB

B1.B.18 "Resistance of materials"

1.Purpose of studying the discipline (module)
The main purpose of studying this discipline, which is an introductory course in deformable solid mechanics for engineers, is: to ensure that future bachelors learn the most important hypotheses, concepts, methods, techniques and approaches to the study of strength, stiffness and stability of structures under static and dynamic actions, necessary in the practical activities of a specialist in the design, manufacture and operation of structures of various purposes, technological equipment, tooling and automation; to give

. Discipline Objectives:
- prepare to solve complex engineering problems using the knowledge base of mathematical and natural science disciplines;
- to achieve that students have mastered the skills to obtain, collect, systematize and analyze the initial information for the development of car and automobile designs and systems;

- to prepare for the development of working technical documentation and execution of completed design works;
- prepare to conduct experiments according to the given methodology and analyze their results.

3. place of the discipline in the structure of the educational program of higher education
The discipline belongs to the basic part of Block 1 "Disciplines (Modules)" of the Bachelor's degree program.

4. A graduate who has mastered the discipline shall possess the following competencies:
OPK-3- have readiness to apply fundamental mathematical, natural-scientific and general engineering knowledge in professional activity
OPK-4 ability to combine theory and practice to solve engineering problems

5. The scope of the discipline (indicating the labor intensity of all types of academic
work)

Types of training work	**Total labor intensity (in hours)**		
	Total ZE /hours	**Semester**	
		3	**4**
Total labor intensity of the discipline (module) in ZE/hour	8 ZE/288.	4 ZE/144.	4 ZE/144.
Contact work of students with the teacher by types of training sessions (classroom work), including:	***120,8***	***70,4***	***50,4***
Lectures	48	34	16
Laboratory work	16		16
Practical exercises	50	34	16
Coursework (consultation, defense)	-	-	-
Course project (consultations, defense)	-	-	-
Pre-exam counseling	4	2	2

Contact work at the intermediate control	0,8	0,4	0,4
Independent work of the student (extracurricular work), including:	***167,2***	***73,6***	***93,6***
Working through the training material (self-training)	100	40	60
Course project (preparation)	-	-	-
Coursework (preparation)	-	-	-
Preparation for interim certification	67,2	33,6	33,6
Intermediate certification		examination	examination

Developer of the RPA: I.N. Sidorov, Ph.D., Professor, Department of TIPM&M, A.I. Kalashnikov, Ph.D., Associate Professor, Department of PC.

B1.B.19 Electrical Engineering and Electronics

1.Purpose of studying the discipline (module)
The purpose of teaching the discipline - to provide basic information on the analysis and synthesis of electric and magnetic circuits; - to study the physical principles, basic characteristics and switching circuits of modern electrical measuring devices, electrical machines and apparatuses, elements of analog and digital electronics

2. Objectives of the discipline:
- to give students knowledge of physical processes and phenomena occurring in DC and AC electrical circuits;
- to give students knowledge about calculation of electric and magnetic circuits using modern mathematical apparatus and computer technology;
- to give students knowledge of the purpose, principle of operation and basic characteristics of single- and three-phase transformers, synchronous and asynchronous machines of alternating current and collector machines of direct current;
- to give students knowledge about the purpose, principle of operation and basic characteristics of elements and units of electronics: transistors, diodes, thyristors, capacitors, resistors, operational amplifiers, comparators, logic elements.
- to give students the knowledge to draw up schematics of the basic nodes of analog electronics (amplifier stages, stabilizers, key circuits, etc.) and simple combinational logic circuits.

3. place of the discipline in the structure of the educational program of higher education

The discipline belongs to the basic part of Block 1 "Disciplines (Modules)" of the Bachelor's degree program.

4. A graduate who has mastered the discipline shall possess the following competencies:

OK-9 readiness to use basic methods of protection of production personnel and population from possible consequences of accidents, catastrophes, natural disasters

OPK-3- have readiness to apply fundamental mathematical, natural-scientific and general engineering knowledge in professional activity

5. The scope of the discipline (indicating the labor intensity of all types of academic work)

Types of training work	**Total labor intensity (in hours)**	
	Total ZE /hours	**Semester**
		5
Total labor intensity of the discipline (module) in ZE/hour	**4/144**	**144**
Contact work of students with the teacher by types of training sessions (classroom work), including:	50,4	50,4
Lectures	34	34
Laboratory work	-	-
Practical exercises	16	16
Coursework (consultation, defense)	-	-
Course project (consultations, defense)	-	-
Pre-exam counseling	2,0	2,0
Contact work at the intermediate control	0,4	0,4
Independent work of the student (extracurricular work), including:	93,6	93,6
Working through the training material (self-training)	60	60
Course project (preparation)	-	-
Coursework (preparation)	-	-
Preparation for interim certification	33,6	33,6

Intermediate certification	Exam	Exam

Developer of RPA: Nizameev I.R., Associate Professor, Department of STEVE, KNITU-KAI, Shakirzyanova N.S., Candidate of Technical Sciences, Associate Professor, Candidate of Technical Sciences, Associate Professor, Department of EE.

B1.B.01 "Physical Education and Sport" (elective discipline)

1.Purpose of studying the discipline (module)
The purpose of physical education of KNITU-KAI students is the formation of physical culture of personality, preservation and promotion of health, psychophysical preparation for future social and professional activities, inclusion in a healthy lifestyle, systematic physical self-improvement.
2. Objectives of the discipline:
1. to determine the role of physical culture in the development of personality and its preparation for socio-professional activity;
2. to study the scientific and practical foundations of physical education and healthy lifestyles;
3. To form students' motivational and value attitude to physical culture, attitude to a healthy lifestyle, physical self-improvement and self-education, the need for regular physical exercise and sports;
4. Acquire practical skills and abilities that ensure the preservation and promotion of health, mental well-being, development and improvement of psychophysical abilities and personality traits, self-determination in physical culture;
5. To provide general and professional-applied physical preparedness, which determines the psychophysical readiness of students for their future profession;
6. Gain experience in creative use of physical culture and sports activities to achieve life and professional goals.

3. place of the discipline in the structure of the educational program of higher education
The discipline "Physical Culture and Sport" (elective discipline) is a continuation of the discipline "Physical Culture and Sport".
It is aimed at successful mastering of other academic disciplines, solving educational, developmental, upbringing and health-improving tasks, ensuring all-round preparedness of the personality.
4. A graduate who has mastered the discipline shall possess the following competencies:

OK - 8 Ability to use methods and means of physical culture to ensure full-fledged social and professional activity

5. The scope of the discipline (indicating the labor intensity of all types of academic
work)

Types of training work	**Total labor intensity (in hours)**					
	Total ZE /hours	**Semester**				
		1	**2**	**3**	**4**	**5**
Total labor intensity of the discipline (module) in ZE/hour	**328**	54	72	72	72	58
Contact work of students with the teacher by types of training sessions (classroom work), including:						
Lectures						
Laboratory work						
Practical exercises	289,5	50	68	68	68	34
Coursework (consultation, defense)	-	-	-	-		
Course project (consultations, defense)	-	-	-	-		
Pre-exam counseling						
Contact work at the intermediate control						
Independent work of the student (extracurricular work), including:						
Working through the training material (self-training)	38,5	3,7	3,7	3,7	3,7	23,7
Course project (preparation)	-	-	-	-		

Coursework (preparation)	-	-	-	-		
Preparation for interim certification	1,5	0,3	0,3	0,3	0,3	0,3
Intermediate certification		credit	credit	credit	cred it	cred it

Developer of the RPA: E.R. Galimov, Ph. E.R. Galimov, Ph.D., Professor, Department of MS&PB, R.A. Yusupov, Ph.

B1.B.02 "Introduction to Professional Activity"

1.Purpose of studying the discipline (module)
The aim of training is to form students' knowledge about the basics of materials science and fields of activity of a full-time bachelor in accordance with the profile "Materials Science and Technology of Materials", including the consideration of the development of materials science in the historical context, revealing, through the specificity of the object and subject of study the individuality of this scientific and academic discipline, objects, areas and types of professional activities of the graduate.

2. Objectives of the discipline:
- familiarization with the history and main stages of development of the science of materials;
- familiarization with modern theories of structure of non-metallic and metallic materials, methods of studying their structure and properties;
- familiarization with basic concepts, terms and definitions in materials science;
- familiarization with the main groups of traditional and modern materials of various functional purposes used in modern technology;
- Familiarization with the basic parameters used to evaluate material properties;
- familiarization with the main directions of regulating the structure and complex of technical properties of materials to create promising materials with specified properties.

3. place of the discipline in the structure of the educational program of higher education
The discipline "Introduction to Professional Activity" is a part of the Variative part of the block B1 "Disciplines (Modules)" of the educational program of Bachelor's degree.

4. A graduate who has mastered the discipline shall possess the following competencies:
PC-2- ability to collect data, study, analyze and summarize scientific and technical information on the subject of research, development and use of technical documentation, basic normative documents on intellectual property issues, preparation of documents for patenting, registration of know-how.
PC-8- readiness to fulfill the basic requirements of office management in relation to records and protocols; to execute project and working technical documentation in accordance with normative documents

5. The scope of the discipline (indicating the labor intensity of all types of academic
work)

Types of training work	**Total labor intensity (in hours)**		
	Total ZE /hours	**Semester**	
		1	
Total labor intensity of the discipline (module) in ZE/hour	2ZE/72.	2ZE/72.	
Contact work of students with the teacher by types of training sessions (classroom work), including:	***16,3***	***16,3***	
Lectures	16	16	
Laboratory work	-	-	
Practical exercises	-	-	
Coursework (consultation, defense)	-	-	
Course project (consultations, defense)	-	-	
Pre-exam counseling	-	-	
Contact work at the intermediate control	0,3	0,3	
Independent work of the student (extracurricular work), including:	***55,7***	***55,7***	
Working through the training material (self-training)	55,7	55,7	
Course project (preparation)	-	-	

Coursework (preparation)	-	-	
Preparation for interim certification	-	-	
Intermediate certification		credit	

Developer of the RPA: E.R. Galimov, Ph. E.R. Galimov, Ph.D., Professor, Department of MS&PS, and P.B. Shibaev, Ph.

B1.B.03 "Materials Science. Technology of structural materials"

1.Purpose of studying the discipline (module)
The aim of training is to form students' fundamental knowledge about the nature and properties of materials, about the dependencies of their properties on composition and structure, about the regularities of transformations in metals and alloys in various thermophysical conditions and processes occurring in materials under load to form skills of scientifically justified choice of materials, application of highly effective methods of their processing and purposeful use in structures with a high degree of reliability and durability.
2. Objectives of the discipline:
- Study of the physical essence of phenomena occurring in materials at the stages of formation of structure and properties, including thermodynamic conditions of transformations and behavior of metals and alloys under load;

- Study of the theory of alloy structure, methods of studying the structure and diagrams of state of alloys;

- Know the basic parameters used to evaluate the properties of modern materials in order to make an informed choice of materials for a given purpose;

- To familiarize with the prospects of creation and use of new materials in connection with the most important directions of development of basic industries;

- Know the patterns of composition, structure and properties of materials.

- Study of modern methods of forming blanks and parts from various materials.

- To get acquainted with the basics of technological processes of casting, OMD, welding, metal forming and other processes of processing and treatment of structural materials, providing high reliability and durability of equipment. Place of the discipline in the structure of the educational program: the discipline "Materials Science. Technology of structural materials" is a part of the variative cycle.

3. place of the discipline in the structure of the educational program of higher education
The discipline belongs to the variable part formed by participants of educational relations, Block 1. Disciplines (modules) of the educational program and is an elective discipline, deepening the mastering of the profile

4. A graduate who has mastered the discipline shall possess the following competencies:
PC-5- readiness to perform complex research and testing in the study of materials and products, including standard and certification, processes of their production, processing and modification.
PC-9- readiness to participate in the development of technological processes of production and processing of coatings, materials and products from them, technological process control systems

5. The scope of the discipline (indicating the labor intensity of all types of academic
work)

Types of training work	**Total labor intensity (in hours)**		
	Total ZE /hours	**Semester**	
		3	**4**
Total labor intensity of the discipline (module) in ZE/hour	8 ZE/288.	4 ZE/144.	4 ZE/144.
Contact work of students with the teacher by types of training sessions (classroom work), including:	***104,8***	***52,4***	***52,4***
Lectures	68	34	34
Laboratory work	-	-	-
Practical exercises	32	16	16
Coursework (consultation, defense)	-	-	-
Course project (consultations, defense)	-	-	-

Pre-exam counseling	4	2	2
Contact work at the intermediate control	0,8	0,4	0,4
Independent work of the student (extracurricular work), including:	***183,2***	***91,6***	***91,6***
Working through the training material (self-training)	116	58	58
Course project (preparation)	-	-	-
Coursework (preparation)	-	-	-
Preparation for interim certification	67,2	33,6	33,6
Intermediate certification		examination	examination

RPA Developer: R.S. Davletbaev, Doctor of Technical Sciences, Professor, Department of MS&PB, F.N. Kurtayeva, Candidate of Technical Sciences, Associate Professor, Department of MS&PB.

B1.V.04 "Physical Chemistry"

1.Purpose of studying the discipline (module)
formation of physical and chemical thinking, skills of theoretical analysis of technological calculations, ability to abstract and build mathematical models of real processes with different degrees of approximation, since any chemical technology is essentially an applied section of physical chemistry.
2. Objectives of the discipline:
to give fundamental concepts and ideas about the theory of chemical processes, a system of general knowledge of the regularities of chemical interaction;
- to teach students to use basic modern physicochemical experimental methods of research and control of chemical processes

3. place of the discipline in the structure of the educational program of higher education
The discipline belongs to the part formed by participants of educational relations, Block 1. Disciplines (modules) of the educational program and is an elective discipline that determines its subject-thematic content and orientation.

4. A graduate who has mastered the discipline shall possess the following competencies:
PC-4- Ability to use in research and calculations knowledge of methods of research, analysis, diagnostics and modeling of properties of substances (materials), physical and chemical processes occurring in materials during their obtaining, processing and modification.

5. The scope of the discipline (indicating the labor intensity of all types of academic
work)

Types of training work	Total labor intensity (in hours)	
	Total ZE /hours	Semester
		2
Total labor intensity of the discipline (module) in ZE/hour	2ZE/72.	72
Contact work of students with the teacher by types of training sessions (classroom work), including:	32,3	32,3
Lectures	16	16
Laboratory work	16	16
Practical exercises	-	-
Coursework (consultation, defense)	-	-
Course project (consultations, defense)	-	-
Pre-exam counseling	-	-
Contact work at the intermediate control	0,3	0,3
Independent work of the student (extracurricular work), including:	39,7	39,7
Working through the training material (self-training)	39,7	39,7
Course project (preparation)	-	-
Coursework (preparation)	-	-
Preparation for interim certification	-	-
Intermediate certification	credit	credit

Developer of RPA: V.A. Morozov, Associate Professor, Department of Chemical Technology and New Materials, Lomonosov Moscow State University, I.G. Grigorieva, Associate Professor, Department of Chemical Technology and New Materials, KNITU-KAI. Department of Chemical Technology and New Materials, KNITU-KAI

B1.V.05 "Organic Chemistry"

1.Purpose of studying the discipline (module)
The purpose of mastering the discipline "Organic Chemistry" is the formation of students fundamental knowledge about the nature and properties of polymeric organic materials used for multi-purpose tasks, about the physical and chemical processes occurring in the materials during their production, processing and modification; ability to use them in research and calculations; ability to choose materials taking into account the requirements of reliability, environmental friendliness.
2. Objectives of the discipline:
- Organic classification
- be able to identify organic substances by their structure;
- study of the regularities of the functional relationship "structure - property";
- To study the chemical principles underlying technological processes;
- be able to write chemical reactions of substances of organic nature;
- Know the methods of creating and modifying materials;
- to make an optimal choice of chemical raw materials for the production of materials according to their properties and purpose;
- ability to apply theoretical knowledge in professional and practical activities of a specialist.
3. place of the discipline in the structure of the educational program of higher education
The discipline belongs to the part formed by participants of educational relations, Block 1. Disciplines (modules) of the educational program and is an elective discipline that determines its subject-thematic content and orientation.

4. A graduate who has mastered the discipline shall possess the following competencies:

PC-4- Ability to use in research and calculations knowledge of methods of research, analysis, diagnostics and modeling of properties of substances (materials), physical and chemical processes occurring in materials during their obtaining, processing and modification.
PC-11- ability to apply knowledge of the main types of modern inorganic and organic materials, principles of materials selection for given operating conditions taking into account the requirements of manufacturability, cost-effectiveness, reliability and durability, environmental consequences of their application in the design of high-tech processes

5. The scope of the discipline (indicating the labor intensity of all types of academic
work)

Types of training work	Total labor intensity (in hours)	
	Total ZE /hours	Semester 5
Total labor intensity of the discipline (module) in ZE/hour	2 ZE/72.	2 ZE/72.
Contact work of students with the teacher by types of training sessions (classroom work), including:	*32,3*	*32,3*
Lectures	16	16
Laboratory work	-	-
Practical exercises	16	16
Coursework (consultation, defense)	-	-
Course project (consultations, defense)	-	-
Pre-exam counseling	-	-
Contact work at the intermediate control	0,3	0,3
Independent work of the student (extracurricular work), including:	*39,7*	*39,7*
Working through the training material (self-training)	39,7	39,7
Course project (preparation)	-	-
Coursework (preparation)	-	-

Preparation for interim certification	-	-
Intermediate certification		credit

Developer of RPA: E.R. Galimov, Doctor of Technical Sciences, Professor, Department of MS&PS of KNITU-KAI, V.I. Treskova, Candidate of Chemical Sciences.

B1.V.06 "Analytical Chemistry"

1.Purpose of studying the discipline (module)
The purpose of studying the discipline "Analytical Chemistry" is the acquisition and formation of knowledge, skills and abilities of future bachelors in the field of application of methods, means and general methodology of obtaining information on the composition and nature of substances
2. Objectives of the discipline:
-Form an understanding of the basic concepts of analytical chemistry and its place in the system of sciences;
-Learn methods for identifying substances (qualitative analysis);
-Learn the methods of quantitative analysis: chemical, physicochemical and physical analysis;
-Learn the metrological foundations of analysis.

3. place of the discipline in the structure of the educational program of higher education
The discipline "Analytical Chemistry" belongs to the variative part of the cycle B.1 disciplines (modules) of the curriculum, is an elective discipline that determines its subject-thematic content - orientation.

4. A graduate who has mastered the discipline shall possess the following competencies:
PC-4- Ability to use in research and calculations knowledge of methods of research, analysis, diagnostics and modeling of properties of substances (materials), physical and chemical processes occurring in materials during their obtaining, processing and modification.

5. The scope of the discipline (indicating the labor intensity of all types of academic
work)

Types of training work	Total labor intensity (in hours)

	Total hours (ZE)	Semester
		5
Total labor intensity of the discipline in ZU/hour	3 ZE/108 .	3 ZE/108 .
Contact work of students with the teacher by types of training sessions (classroom work), including:	*34,4*	*34,4*
Lectures	16	16
Laboratory work	16	16
Practical exercises	-	-
Contact work at the intermediate control	2,4	2,4
Independent work of the student (extracurricular work), including:	*73,6*	*73,6*
Working through the training material (self-training)	40	40
Preparation for interim certification	33,6	33,6
Intermediate certification		examination

RPA Developer: R.S. Davletbaev, Doctor of Chemical Sciences, Professor of the Department of MS&PB

B1.V.07 "Physical Chemistry of Materials"

1.Purpose of studying the discipline (module)
The purpose of studying the discipline is to form students' knowledge about the regularities of influence of chemical composition, structure, phase and physical states of substances on the complex of operational properties of products based on them.
2. Objectives of the discipline:
-The main objectives of mastering the discipline are:
-In-depth theoretical and practical mastering of the basic concepts of composition and structure, geometric and phase structure of inorganic and organic matrix and reinforcing materials at the level of atoms, bonds, molecules, atomic and molecular lattices, amorphous and crystalline phases;
-forming knowledge of the principles of directed creation and regulation of their phase structure and interaction of components and phases at the interface;

-analysis of the influence of the nature and properties of components (phases), their volume fractions and character of distribution (phase structure), as well as the interaction at the interface on the basic physical-chemical and physical-mechanical properties and other operational properties of materials.

3. place of the discipline in the structure of the educational program of higher education
The discipline "Physical Chemistry of Materials" belongs to the variative part of the cycle B.1 disciplines (modules) of the curriculum, is an elective discipline that determines its subject-thematic content - orientation.

4. A graduate who has mastered the discipline shall possess the following competencies:
PC-4- Ability to use in research and calculations knowledge of methods of research, analysis, diagnostics and modeling of properties of substances (materials), physical and chemical processes occurring in materials during their obtaining, processing and modification.
PC-6 Ability to use in practice modern ideas about the influence of microstructure and nanostructure on the properties of materials, their interaction with the environment, fields, particles and radiation.

5. The scope of the discipline (indicating the labor intensity of all types of academic
work)

Types of training work	Total labor intensity (in hours)		
	Total hours (ZE)	Semester	
		5	7
Total labor intensity of the discipline in ZU/hour	7 ZE/252.	3 ZE/108.	4 ZE/144.
Contact work of students with the teacher by types of training sessions (classroom work), including:	*98,7*	*48,3*	*50,4*
Lectures	32	16	16
Laboratory work	32	16	16
Practical exercises	32	16	16
Pre-exam counseling	2		2
Contact work at the intermediate control	0,7	0,3	0,4

Independent work of the student (extracurricular work), including:	*153,3*	*59,7*	*93,6*
Working through the training material (self-training)		59,7	60
Course project (preparation)	-	-	-
Coursework (preparation)	-	-	-
Preparation for interim certification	33,6		33,6
Intermediate certification			examin ation

R.S. Davletbaev, Doctor of Chemical Sciences, Professor, Department of MS&PS, K.A. Andrianova, Candidate of Technical Sciences, Associate Professor, Department of PLA.

B.1.V.08_"Diagnostics and safety assurance of technological processes and equipment"

1.Purpose of studying the discipline (module)

The purpose of training is to form the future academic bachelors' understanding of the inseparable unity of effective professional activity with the requirements for human safety and security. The realization of these requirements guarantees the preservation of performance of technological processes, equipment and human health safety, prepares him to act in extreme conditions, as well as the formation of skills in the field of using methods of equipment diagnostics and in mastering the issues of industrial safety at enterprises.

2. Objectives of the discipline:

--Research and creation of comfortable (normative) state of habitat in areas of human labor activity and recreation;

-Possession of technologies of condition assessment by diagnostic methods;

-Knowledge of identification of negative impacts of habitat of natural, man-made and anthropogenic origin;

- Be able to predict the development of negative impacts on humans and the environment, risk assessment and management.

- Be able to develop and implement measures to protect humans and the environment from negative impacts;

-To be able to design techniques for diagnostics of technological processes and economic objects in accordance with the requirements for safety and environmental friendliness;

-Mastering of methods for determining the zones of increased technogenic risk, selection of human protection system in the operation of certain types of technological equipment and production processes.
-Securing stability of functioning of facilities and technical systems in regular and emergency situations;
The ability to make decisions on the protection of production personnel and population from the possible consequences of accidents, disasters, natural disasters and the use of modern means of destruction, as well as taking measures to eliminate their consequences; Place of the discipline in the structure of the educational program: the discipline "Diagnostics and ensuring the safety of technological processes and equipment" is part of the variative cycle.
3. place of the discipline in the structure of the educational program of higher education
The discipline belongs to the variative part of the cycle B.1 disciplines (modules) of the curriculum, is an elective discipline that determines its subject-thematic content - orientation.

4. A graduate who has mastered the discipline shall possess the following competencies:
PC-10- ability to assess the quality of materials in production conditions at the stage of pilot testing and implementation.
5. The scope of the discipline (indicating the labor intensity of all types of academic
work)

Types of training work	Total labor intensity (in hours)	
	Total ZE /hours	Semester
		7
Total labor intensity of the discipline (module) in ZE/hour	2 ZE/72.	2 ZE/72.
Contact work of students with the teacher by types of training sessions (classroom work), including:	*16,3*	*16,3*
Lectures	16	16
Laboratory work	-	-
Practical exercises	-	-

Coursework (consultation, defense)	-	-
Course project (consultations, defense)	-	-
Pre-exam counseling	-	-
Contact work at the intermediate control	0,8	0,3
Independent work of the student (extracurricular work), including:	*55,7*	*55,7*
Working through the training material (self-training)	-	-
Course project (preparation)	-	-
Coursework (preparation)	-	-
Preparation for interim certification	-	-
Intermediate certification		credit

RPA Developer: E.R. Galimov, Doctor of Technical Sciences, Professor, Department of MS&PS of KNITU-KAI, V.Y. Vinogradov, Candidate of Technical Sciences, Associate Professor, Department of MS&PS.

B.1.V.08_"Diagnostics and safety assurance of technological processes and equipment"

1.Purpose of studying the discipline (module)

The purpose of training is to form the future academic bachelors' understanding of the inseparable unity of effective professional activity with the requirements for human safety and security. The realization of these requirements guarantees the preservation of performance of technological processes, equipment and human health safety, prepares him to act in extreme conditions, as well as the formation of skills in the field of using methods of equipment diagnostics and in mastering the issues of industrial safety at enterprises.

2. Objectives of the discipline:

--Research and creation of comfortable (normative) state of habitat in areas of human labor activity and recreation;

-Possession of technologies of condition assessment by diagnostic methods;

-Knowledge of identification of negative impacts of habitat of natural, man-made and anthropogenic origin;

- Be able to predict the development of negative impacts on humans and the environment, risk assessment and management.
- Be able to develop and implement measures to protect humans and the environment from negative impacts;

-To be able to design techniques for diagnostics of technological processes and economic objects in accordance with the requirements for safety and environmental friendliness;
-Mastering of methods for determining the zones of increased technogenic risk, selection of human protection system in the operation of certain types of technological equipment and production processes.
-Securing stability of functioning of facilities and technical systems in regular and emergency situations;
The ability to make decisions on the protection of production personnel and population from the possible consequences of accidents, disasters, natural disasters and the use of modern means of destruction, as well as taking measures to eliminate their consequences; Place of the discipline in the structure of the educational program: the discipline "Diagnostics and ensuring the safety of technological processes and equipment" is part of the variative cycle.
3. place of the discipline in the structure of the educational program of higher education
The discipline belongs to the variative part of the cycle B.1 disciplines (modules) of the curriculum, is an elective discipline that determines its subject-thematic content - orientation.

4. A graduate who has mastered the discipline shall possess the following competencies:
PC-10- ability to assess the quality of materials in production conditions at the stage of pilot testing and implementation.
5. The scope of the discipline (indicating the labor intensity of all types of academic
work)

Types of training work	Total labor intensity (in hours)	
	Total ZE /hours	Semester 7
Total labor intensity of the discipline (module) in ZE/hour	2 ZE/72.	2 ZE/72.
Contact work of students with the teacher by types of training sessions (classroom work), including:	*16,3*	*16,3*
Lectures	16	16

Laboratory work	-	-
Practical exercises	-	-
Coursework (consultation, defense)	-	-
Course project (consultations, defense)	-	-
Pre-exam counseling	-	-
Contact work at the intermediate control	0,8	0,3
Independent work of the student (extracurricular work), including:	*55,7*	*55,7*
Working through the training material (self-training)	-	-
Course project (preparation)	-	-
Coursework (preparation)	-	-
Preparation for interim certification	-	-
Intermediate certification		credit

RPA Developer: E.R. Galimov, Doctor of Technical Sciences, Professor, Department of MS&PS of KNITU-KAI, V.Y. Vinogradov, Candidate of Technical Sciences, Associate Professor, Department of MS&PS.

B1.V.DV.02.01 "Alloy Theory"

1.Purpose of studying the discipline (module)
The purpose of training is to form students' fundamental knowledge of the nature, structure and properties of materials, the dependencies of their properties on the composition and structure, the regularities of transformations in metals and alloys in different thermophysical conditions and processes occurring in materials under load to form skills of scientifically-based choice of materials.
2. Objectives of the discipline:
- study the theoretical basis of crystallization of metal alloys;
- To study the actual structure and the role of defects in the crystal structure of metal alloys;
- basic diffusion patterns and structural composition of alloys;
- theoretical bases of elastic and plastic deformation of alloys;
- master various methods of determining the mechanical properties of materials and their certification and incoming inspection;
- to study state diagrams in relation to typical structural alloys of transport and power engineering, to master practical skills of regularities of state

diagram - properties (according to N.S. Kurnakov) to solve problems of substantiation of materials with specified properties.

3. place of the discipline in the structure of the educational program of higher education
The discipline belongs to the variative part of Block 1. Disciplines (modules) of the educational program and is a discipline of choice, deepening the mastering of the profile

4. A graduate who has mastered the discipline shall possess the following competencies:
PC-4 ability to use in research and calculations knowledge of methods of research, analysis, diagnostics and modeling of properties of substances (materials), physical and chemical processes occurring in materials during their obtaining, processing and modification.

PC-5 readiness to perform complex research and tests in the study of materials and products, including standard and certification, their production, processing and modification processes
5. The scope of the discipline (indicating the labor intensity of all types of academic
work)

Types of training work	Total labor intensity (per hour)	
	Total hours (ZE)	Semester 4
Total labor intensity of the discipline (module) in ZE/hour	3 ZE/108.	3 ZE/108.
Contact work of students with the teacher by types of training sessions (classroom work), including:	*32,3*	*32,3*
Lectures	16	16
Laboratory work	16	16
Practical exercises	-	-
Coursework (consultation, defense)	-	-
Course project (consultations, defense)	-	-
Pre-exam counseling	-	-
Contact work at the intermediate control	0,3	0,3

Independent work of the student (extracurricular work), including:	*75,7*	*75,7*
Working through the training material (self-training)	75,7	75,7
Course project (preparation)	-	-
Coursework (preparation)	-	-
Preparation for interim certification	-	-
Intermediate certification	credit	credit

RPA developer: E.R. Galimov, Doctor of Technical Sciences, Professor, Department of MS&PS of KNITU-KAI, F.I. Murataev, Candidate of Technical Sciences, Associate Professor, Department of MS&PS.

B1.V.DV.02.02 "Theoretical Foundations of Metallurgy"

1.Purpose of studying the discipline (module)
The purpose of training is to form students' fundamental knowledge of the nature, structure and properties of materials, the dependencies of their properties on the composition and structure, the regularities of transformations in metals and alloys in different thermophysical conditions and processes occurring in materials under load to form skills of scientifically-based choice of materials.
2. Objectives of the discipline:
- study the theoretical basis of crystallization of metal alloys;
- To study the actual structure and the role of defects in the crystal structure of metal alloys;
- basic diffusion patterns and structural composition of alloys;
- theoretical bases of elastic and plastic deformation of alloys;
- master various methods of determining the mechanical properties of materials and their certification and incoming inspection;
- to study state diagrams in relation to typical structural alloys of transport and power engineering, to master practical skills of regularities of state diagram - properties (according to N.S. Kurnakov) to solve problems of substantiation of materials with specified properties.

3. place of the discipline in the structure of the educational program of higher education
The discipline belongs to the variative part of Block 1. Disciplines (modules) of the educational program and is a discipline of choice, deepening the mastering of the profile

4. A graduate who has mastered the discipline shall possess the following competencies:
PC-4 ability to use in research and calculations knowledge of methods of research, analysis, diagnostics and modeling of properties of substances (materials), physical and chemical processes occurring in materials during their obtaining, processing and modification.
PC-5 readiness to perform complex research and tests in the study of materials and products, including standard and certification, their production, processing and modification processes
5. The scope of the discipline (indicating the labor intensity of all types of academic
work)

Types of training work	Total labor intensity (per hour)	
	Total hours (ZE)	Semester 4
Total labor intensity of the discipline (module) in ZE/hour	3 ZE/108.	3 ZE/108.
Contact work of students with the teacher by types of training sessions (classroom work), including:	*32,3*	*32,3*
Lectures	16	16
Laboratory work	16	16
Practical exercises	-	-
Coursework (consultation, defense)	-	-
Course project (consultations, defense)	-	-
Pre-exam counseling	-	-
Contact work at the intermediate control	0,3	0,3
Independent work of the student (extracurricular work), including:	*75,7*	*75,7*
Working through the training material (self-training)	75,7	75,7
Course project (preparation)	-	-
Coursework (preparation)	-	-
Preparation for interim certification	-	-
Intermediate certification	credit	credit

RPA developer: E.R. Galimov, Doctor of Technical Sciences, Professor, Department of MS&PS of KNITU-KAI, F.I. Murataev, Candidate of Technical Sciences, Associate Professor, Department of MS&PS.

B1.V.DV.03.01 "Physical bases of non-metallic materials research".

1.Purpose of studying the discipline (module)
The main purpose of the discipline is to familiarize students with the current level of development of research techniques and technology, the capabilities of various research methods, with their equipment and conditions of experimentation; formation of skills of comparative assessment of the capabilities of different methods of analysis, their advantages and disadvantages for a reasonable choice of the optimal method of research of this or that object....
2. Objectives of the discipline:
-Study of the physical theory of materials research methods, experimentation schemes and techniques;
-Formation of ideas about the possibilities of using certain physical research methods to solve inverse problems, i.e. to determine the required parameters of research objects;
-Analysis of the capabilities of modern physical research methods in terms of their theoretical and practical applications, including industrial applications.

3. place of the discipline in the structure of the educational program of higher education
The discipline belongs to the variative part of Block 1. Disciplines (modules) of the educational program and is a discipline of choice, deepening the mastering of the profile

4. A graduate who has mastered the discipline shall possess the following competencies:
PC-6 Ability to use in practice modern ideas about the influence of microstructure and nanostructure on the properties of materials, their interaction with the environment, fields, particles and radiation.

5. The scope of the discipline (indicating the labor intensity of all types of academic
work)

Types of training work	Total labor intensity (in hours)	
	Total hours (ZE)	Semester
		5
Total labor intensity of the discipline in ZU/hour	3 ZE/108 .	3 ZE/108 .
Contact work of students with the teacher by types of training sessions (classroom work), including:	*48,3*	*48,3*
Lectures	16	16
Laboratory work	32	32
Practical exercises	-	-
Contact work at the intermediate control	0,3	0,3
Independent work of the student (extracurricular work), including:	*59,7*	*59,7*
Working through the training material (self-training)	59,7	59,7
Preparation for interim certification		
Intermediate certification		credit

RPA Developer: R.S. Davletbaev, Doctor of Chemical Sciences, Professor, Department of MS&PS of KNITU-KAI

B1.B.ДВ.03.02 "Methods of research of polymeric materials"

1.Purpose of studying the discipline (module)
The main purpose of the discipline is to familiarize students with the current level of development of research techniques and technology, the capabilities of various research methods, with their equipment and conditions of experimentation; formation of skills of comparative assessment of the capabilities of different methods of analysis, their advantages and disadvantages for a reasonable choice of the optimal method of research of this or that object....
2. Objectives of the discipline:

-Study of the physical theory of materials research methods, experimentation schemes and techniques;
-Formation of ideas about the possibilities of using certain physical research methods to solve inverse problems, i.e. to determine the required parameters of research objects;
-Analysis of the capabilities of modern physical research methods in terms of their theoretical and practical applications, including industrial applications.
3. place of the discipline in the structure of the educational program of higher education
The discipline belongs to the variative part of Block 1. Disciplines (modules) of the educational program and is a discipline of choice, deepening the mastering of the profile
4. A graduate who has mastered the discipline shall possess the following competencies:
PC-6 Ability to use in practice modern ideas about the influence of microstructure and nanostructure on the properties of materials, their interaction with the environment, fields, particles and radiation.
5. The scope of the discipline (indicating the labor intensity of all types of academic
work)

Types of training work	Total labor intensity (in hours)	
	Total hours (ZE)	Semester 5
Total labor intensity of the discipline in ZU/hour	3 ZE/108 .	3 ZE/108 .
Contact work of students with the teacher by types of training sessions (classroom work), including:	*48,3*	*48,3*
Lectures	16	16
Laboratory work	32	32
Practical exercises	-	-
Contact work at the intermediate control	0,3	0,3
Independent work of the student (extracurricular work), including:	*59,7*	*59,7*
Working through the training material (self-training)	59,7	59,7
Preparation for interim certification		

Intermediate certification		credit

RPA Developer: R.S. Davletbaev, Doctor of Chemical Sciences, Professor, Department of MS&PS of KNITU-KAI

B1.V.DV.04.01 "Physical bases of metallic materials research"

1.Purpose of studying the discipline (module)

The purpose of studying the discipline is to form future bachelors' knowledge of classical and advanced methods of research and determination of the nature of metallic materials and their properties, abilities and skills to apply natural science and engineering knowledge in the study of materials at the micro- and nano-level.

2. Objectives of the discipline:

- study of physical bases of methods of research, diagnostics, testing of metallic materials;
- The ability to select the type, method or technique of examination of metallic material, depending on the nature of the material and the object of examination;
- Study of advanced and modern techniques for the study of metallic materials.

3. place of the discipline in the structure of the educational program of higher education

The discipline belongs to the variative part of Block 1. Disciplines (modules) of the educational program and is a discipline of choice, deepening the mastering of the profile

4. A graduate who has mastered the discipline shall possess the following competencies:

PC-3 readiness to use modeling methods in predicting and optimizing technological processes and properties of materials, standardization and certification of materials and processes

PC-6 ability to use in practice modern ideas about the influence of microstructure and nanostructure on the properties of materials, their interaction with the environment, fields, particles and radiation.

PC-14 readiness to use technical means of measurement and control required for standardization and certification of materials and processes of their production, testing and production equipment

5. The scope of the discipline (indicating the labor intensity of all types of academic
work)

Types of training work	**Total labor intensity (in hours)**	
	Total ZE /hours	**Semester**
		6
Total labor intensity of the discipline (module) in ZE/hour	4 ZE/144.	4 ZE/144.
Contact work of students with the teacher by types of training sessions (classroom work), including:	34,4	34,4
Lectures	16	16
Laboratory work	16	16
Practical exercises	-	-
Coursework (consultation, defense)	-	-
Course project (consultations, defense)	-	-
Pre-exam counseling	2	2
Contact work at the intermediate control	0,4	0,4
Independent work of the student (extracurricular work), including:	109,6	109,6
Working through the training material (self-training)	76	76
Course project (preparation)	-	-
Coursework (preparation)	-	-
Preparation for interim certification	33,6	33,6
Intermediate certification		examination

RPA Developer: S.V. Kuryntsev, Candidate of Economic Sciences, Associate Professor, Department of MS&PB, KNITU-KAI

B1.V.DV.04.02 "Methods of research of metallic materials"
1.Purpose of studying the discipline (module)
The purpose of studying the discipline is to form future bachelors' knowledge of classical and advanced methods of research and determination of the nature of metallic materials and their properties, abilities and skills to apply natural science and engineering knowledge in the study of materials at the micro- and nano-level.
2. Objectives of the discipline:

- study of physical bases of methods of research, diagnostics, testing of metallic materials;
- The ability to select the type, method or technique of examination of metallic material, depending on the nature of the material and the object of examination;
- Study of advanced and modern techniques for the study of metallic materials.

3. place of the discipline in the structure of the educational program of higher education

The discipline belongs to the variative part of Block 1. Disciplines (modules) of the educational program and is a discipline of choice, deepening the mastering of the profile

4. A graduate who has mastered the discipline shall possess the following competencies:

PC-3 readiness to use modeling methods in predicting and optimizing technological processes and properties of materials, standardization and certification of materials and processes

PC-6 ability to use in practice modern ideas about the influence of microstructure and nanostructure on the properties of materials, their interaction with the environment, fields, particles and radiation.

PC-14 readiness to use technical means of measurement and control required for standardization and certification of materials and processes of their production, testing and production equipment

5. The scope of the discipline (indicating the labor intensity of all types of academic
work)

Types of training work	Total labor intensity (in hours)	
	Total ZE /hours	Semester 6
Total labor intensity of the discipline (module) in ZE/hour	4 ZE/144.	4 ZE/144.
Contact work of students with the teacher by types of training sessions (classroom work), including:	34,4	34,4
Lectures	16	16
Laboratory work	16	16
Practical exercises	-	-
Coursework (consultation, defense)	-	-

Course project (consultations, defense)	-	-
Pre-exam counseling	2	2
Contact work at the intermediate control	0,4	0,4
Independent work of the student (extracurricular work), including:	109,6	109,6
Working through the training material (self-training)	76	76
Course project (preparation)	-	-
Coursework (preparation)	-	-
Preparation for interim certification	33,6	33,6
Intermediate certification		examination

RPA Developer: S.V. Kuryntsev, Candidate of Economic Sciences, Associate Professor, Department of MS&PB, KNITU-KAI

B1.B.ДB.05.01 "Metallographic analysis in production conditions"

1.Purpose of studying the discipline (module)

The main purpose of studying the discipline is the formation of skills and abilities of experimental study of the structure and properties of materials, to use in practice modern ideas about the influence of microstructure and nanostructure on the properties of materials.

2. Objectives of the discipline:

- mastering by students the knowledge of the real structure of metallic materials, its connection with phase equilibrium diagrams and types of heat treatment;
- mastering by students the methods of light microscopy at macro- and meso-level;
- mastering of metallographic analysis of carbon and alloy steels of different structural classes;
- mastering fracture analysis to determine the causes of structural failure.
- mastering of basic concepts and terms related to electron microscopy;
- familiarization with the device of a standard electron microscope
- study of the principles of robotic electronic optics.

3. place of the discipline in the structure of the educational program of higher education

The discipline belongs to the variative part of Block 1. Disciplines (modules) of the educational program and is a discipline of choice, deepening the mastering of the profile

4. A graduate who has mastered the discipline shall possess the following competencies:

PC 6- ability to use in practice modern ideas about the influence of micro- and nano-structure on the properties of materials, their interaction with the environment, fields, particles and radiation.

PC-14 - readiness to use technical means of measurement and control required for standardization and certification of materials and processes of their production, testing and production equipment.

5. The scope of the discipline (indicating the labor intensity of all types of academic
work)

Types of training work	**Total labor intensity (in hours)**		
	Total ZE /hours	**Semester**	
		5	**6**
Total labor intensity of the discipline (module) in ZE/hour	7 ZE/252.	3 ZE/108.	4 ZE/144.
Contact work of students with the teacher by types of training sessions (classroom work), including:	***82,7***	***32,3***	***52,4***
Lectures	48	16	34
Laboratory work	32	16	16
Practical exercises	-	-	-
Coursework (consultation, defense)	-	-	-
Course project (consultations, defense)	-	-	-
Pre-exam counseling	2	-	2
Contact work at the intermediate control	0,7	0,3	0,4
Independent work of the student (extracurricular work), including:	***135,7***	***75,7***	***91,6***
Working through the training material (self-training)	135,7	75,7	58
Course project (preparation)	-	-	-
Coursework (preparation)	-	-	-
Preparation for interim certification	33,6	-	33,6
Intermediate certification		credit	examination

Developer of the RPA: Nizameev I.R. I.R. Nizameev, Candidate of Chemical Sciences, Associate Professor, Department of STEVE, F.N. Kurtayeva, Candidate of Technical Sciences, Associate Professor, Department of MSIPB, KNITU-KAI.

B1.V.DV.05.01 "Metallographic analysis in scientific research"

1.Purpose of studying the discipline (module)

The main purpose of studying the discipline is the formation of skills and abilities of experimental study of the structure and properties of materials, to use in practice modern ideas about the influence of microstructure and nanostructure on the properties of materials.

2. Objectives of the discipline:

- mastering by students the knowledge of the real structure of metallic materials, its connection with phase equilibrium diagrams and types of heat treatment;
- mastering by students the methods of light microscopy at macro- and meso-level;
- mastering of metallographic analysis of carbon and alloy steels of different structural classes;
- mastering fracture analysis to determine the causes of structural failure.
- mastering of basic concepts and terms related to electron microscopy;
- familiarization with the device of a standard electron microscope
- study of the principles of robotic electronic optics.

3. place of the discipline in the structure of the educational program of higher education

The discipline belongs to the variant part of Block 1. Disciplines (modules) of the educational program and is a discipline of choice, deepening the mastering of the profile

4. A graduate who has mastered the discipline shall possess the following competencies:

PC 6- ability to use in practice modern ideas about the influence of micro- and nano-structure on the properties of materials, their interaction with the environment, fields, particles and radiation.

PC-14 - readiness to use technical means of measurement and control required for standardization and certification of materials and processes of their obtaining, testing and production equipment.

5. The scope of the discipline (indicating the labor intensity of all types of academic work)

Types of training work	**Total labor intensity (in hours)**		
	Total ZE /hours	**Semester**	
		5	**6**
Total labor intensity of the discipline (module) in ZE/hour	7 ZE/252.	3 ZE/108.	4 ZE/144.

Contact work of students with the teacher by types of training sessions (classroom work), including:	***82,7***	***32,3***	***52,4***
Lectures	48	16	34
Laboratory work	32	16	16
Practical exercises	-	-	-
Coursework (consultation, defense)	-	-	-
Course project (consultations, defense)	-	-	-
Pre-exam counseling	2	-	2
Contact work at the intermediate control	0,7	0,3	0,4
Independent work of the student (extracurricular work), including:	***135,7***	*75,7*	***91,6***
Working through the training material (self-training)	135,7	75,7	58
Course project (preparation)	-	-	-
Coursework (preparation)	-	-	-
Preparation for interim certification	33,6	-	33,6
Intermediate certification		credit	examination

Developer of RPA: Nizameev I.R. I.R. Nizameev, Candidate of Chemical Sciences, Associate Professor, Department of STEVE,

B1.V.DV.06.01 "Mathematical modeling of vibration state of technical systems"

1.Purpose of studying the discipline (module)

The aim of the course is to form students' knowledge, skills and abilities to use methods of modeling of vibration state of rotor systems taking into account the properties of materials, individual components of the system, as well as building models to predict the results of the technological process of balancing.

2. Objectives of the discipline:

- gaining knowledge of balancing technical systems;
- study of typical technological processes of balancing flexible rotors;
- obtaining skills of calculation of equations of the mathematical model of the technical system;
- Acquisition of skills and abilities to work in software tools for balancing technical systems;
- acquisition of skills and abilities to think logically, to formulate mathematical models and set problems, to analyze equations and construct solutions, to apply the acquired knowledge to solve actual practical problems.

3. place of the discipline in the structure of the educational program of higher education

The discipline belongs to the variative part of Block 1. Disciplines (modules) of the educational program and is a discipline of choice, deepening the mastering of the profile

4. A graduate who has mastered the discipline shall possess the following competencies:

PC-3 - readiness to use modeling methods in prediction and optimization of technological processes and properties of materials, standardization and certification of materials and processes

PC-7 - ability to select and apply appropriate methods of modeling physical, chemical and technological processes

5. The scope of the discipline (indicating the labor intensity of all types of academic work)

Types of training work	**Total labor intensity (in hours)**	
	Total ZE /hours	**Semester 6**
Total labor intensity of the discipline (module) in ZE/hour	2 ZE/72.	2 ZE/72.
Contact work of students with the teacher by types of training sessions (classroom work), including:	***32,3***	***32,3***
Lectures	16	16
Laboratory work	-	-
Practical exercises	16	16
Coursework (consultation, defense)	-	-
Course project (consultations, defense)	-	-
Pre-exam counseling	-	-
Contact work at the intermediate control	0,3	0,3
Independent work of the student (extracurricular work), including:	***39,7***	***39,7***
Working through the training material (self-training)	39,7	39,7
Course project (preparation)	-	-
Coursework (preparation)	-	-
Preparation for interim certification	-	-
Intermediate certification		credit

Developer of RPA: RPA: I.N. Sidorov, Doctor of Physics and Mathematics, Professor, Department of TIPM&M.

B1.B.ДВ.06.02 "Mathematical modeling in problems of mechanics of composite materials"

1.Purpose of studying the discipline (module)

The aim of the course is to form knowledge, skills and abilities of future bachelors in modeling the processes of deformation of elements of composite structures under external loading. Ensuring the students' mastering of the most important hypotheses, concepts, methods, techniques and approaches to the study of the deformation process of structural elements, necessary in practical activities in the design of aviation and other types of structures.

2. Objectives of the discipline:

The main objectives of the discipline are:

- development of skills of mathematical and mechanical approaches to the problem of modeling of various physical phenomena of mechanics of composite materials;
- Ability to think logically, formulate mathematical models and problem statements, analyze equations and construct solutions, and apply acquired knowledge to solve actual practical problems

3. place of the discipline in the structure of the educational program of higher education

The discipline belongs to the variative part of Block 1. Disciplines (modules) of the educational program and is a discipline of choice, deepening the mastering of the profile

4. A graduate who has mastered the discipline shall possess the following competencies:

PC-3 - readiness to use modeling methods in prediction and optimization of technological processes and properties of materials, standardization and certification of materials and processes

PC-7 - ability to select and apply appropriate methods of modeling physical, chemical and technological processes

5. The scope of the discipline (indicating the labor intensity of all types of academic
work)

Types of training work	**Total labor intensity (in hours)**	
		Semester

	Total ZE /hours	**6**
Total labor intensity of the discipline (module) in ZE/hour	2 ZE/72.	2 ZE/72.
Contact work of students with the teacher by types of training sessions (classroom work), including:	***32,3***	***32,3***
Lectures	16	16
Laboratory work	-	-
Practical exercises	16	16
Coursework (consultation, defense)	-	-
Course project (consultations, defense)	-	-
Pre-exam counseling	-	-
Contact work at the intermediate control	0,3	0,3
Independent work of the student (extracurricular work), including:	***39,7***	***39,7***
Working through the training material (self-training)	39,7	39,7
Course project (preparation)	-	-
Coursework (preparation)	-	-
Preparation for interim certification	-	-
Intermediate certification		credit

Developer of RPA: RPA: I.N. Sidorov, Doctor of Physics and Mathematics, Professor, Department of TIPMiM.

B1.V.DV.08.01 "Diagnostics, control and quality management of technological processes and materials".

1.Purpose of studying the discipline (module)

The aim of the training is to study the regularities of degradation of composition, structure and properties of materials, taking into account the actual stress-strain state, their operational and technological heredity in the parts of complex structures, machinery and devices, to identify the technical condition or determine the causes of incidents, failures, critical conditions and accidents.

2. Objectives of the discipline:

The main objectives of the discipline are for students to master:

- data bank on structural materials, standard technologies applied to them, regularities of formation of structure and properties of materials and welded joints, conditions for ensuring their stability in operation;
- influence of operating conditions on the properties of materials and the main regularities of degradation of their composition and structure;

- instruments and equipment for destructive and non-destructive testing, for determination and assessment of the quality of materials and welded joints, taking into account the requirements of the current system of conformity assessment and NDT;
- methods of destructive and non-destructive testing, to determine and evaluate the quality of materials and welded joints;
- rejection characteristics of composition, structure and properties of materials and welded joints.

3. place of the discipline in the structure of the educational program of higher education

The discipline belongs to the variative part of Block 1. Disciplines (modules) of the educational program and is a discipline of choice, deepening the mastering of the profile

4. A graduate who has mastered the discipline shall possess the following competencies:

PC-5 readiness to perform complex research and tests in the study of materials and products, including standard and certification, their production, processing and modification processes

PC-10 ability to assess the quality of materials in production conditions at the stage of pilot testing and implementation

5. The scope of the discipline (indicating the labor intensity of all types of academic
work)

Types of training work	**Total labor intensity (in hours)**		
	Total ZE /hours	**Semester**	
		6	**7**
Total labor intensity of the discipline (module) in ZE/hour	8 ZE/288.	4 ZE/144.	4 ZE/144.
Contact work of students with the teacher by types of training sessions (classroom work), including:	***102,7***	***52,4***	***50,3***
Lectures	68	34	34
Laboratory work	32	16	16
Practical exercises	-	-	-
Coursework (consultation, defense)	-	-	-
Course project (consultations, defense)	-	-	-
Pre-exam counseling	2	2	-
Contact work at the intermediate control	0,7	0,4	0,3

Independent work of the student (extracurricular work), including:	*151,7*	*121,6*	*93,7*
Working through the training material (self-training),	118,1	58	93,7
Course project (preparation)	-	-	-
Coursework (preparation)	-	-	-
Preparation for interim certification	33,6	33,6	
Intermediate certification		examination	credit

RPA Developer: E.R. Galimov, Doctor of Technical Sciences, Professor, Department of MS&PS of KNITU-KAI, F.I. Murataev, Candidate of Technical Sciences, Associate Professor, Department of MS&PS.

B1.V.DV.08.02 "Quality assurance of materials in design, production and operation"

1.Purpose of studying the discipline (module)

The aim of the training is to study the regularities of degradation of composition, structure and properties of materials, taking into account the actual stress-strain state, their operational and technological heredity in the parts of complex structures, machinery and devices, to identify the technical condition or determine the causes of incidents, failures, critical conditions and accidents.

2. Objectives of the discipline:

The main objectives of the discipline are for students to master:

- data bank on structural materials, standard technologies applied to them, regularities of formation of structure and properties of materials and welded joints, conditions for ensuring their stability in operation;
- influence of operating conditions on material properties and basic regularities of degradation of their composition and structure;
- instruments and equipment for destructive and non-destructive testing, for determination and assessment of the quality of materials and welded joints, taking into account the requirements of the current system of conformity assessment and NDT;
- methods of destructive and non-destructive testing, to determine and evaluate the quality of materials and welded joints;
- rejection characteristics of composition, structure and properties of materials and welded joints.

3. place of the discipline in the structure of the educational program of higher education

The discipline belongs to the variative part of Block 1. Disciplines (modules) of the educational program and is a discipline of choice, deepening the mastering of the profile

4. A graduate who has mastered the discipline shall possess the following competencies:

PC-5 readiness to perform complex research and tests in the study of materials and products, including standard and certification, their production, processing and modification processes

PC-10 ability to assess the quality of materials in production conditions at the stage of pilot testing and implementation

5. The scope of the discipline (indicating the labor intensity of all types of academic
work)

Types of training work	**Total labor intensity (in hours)**		
	Total ZE /hours	**Semester**	
		6	**7**
Total labor intensity of the discipline (module) in ZE/hour	8 ZE/288.	4 ZE/144.	4 ZE/144.
Contact work of students with the teacher by types of training sessions (classroom work), including:	***102,7***	***52,4***	***50,3***
Lectures	68	34	34
Laboratory work	32	16	16
Practical exercises	-	-	-
Coursework (consultation, defense)	-	-	-
Course project (consultations, defense)	-	-	-
Pre-exam counseling	2	2	-
Contact work at the intermediate control	0,7	0,4	0,3
Independent work of the student (extracurricular work), including:	***151,7***	***121,6***	***93,7***
Working through the training material (self-training),	118,1	58	93,7
Course project (preparation)	-	-	-
Coursework (preparation)	-	-	-
Preparation for interim certification	33,6	33,6	

Intermediate certification		examin ation	credit

RPA Developer: E.R. Galimov, Doctor of Technical Sciences, Professor, Department of MS&PS of KNITU-KAI, F.I. Murataev, Candidate of Technical Sciences, Associate Professor, Department of MS&PS.

B1.V.DV.09.01 "Technological preparation of production"

1.Purpose of studying the discipline (module)

The purpose of studying the discipline is to prepare a bachelor to develop measures for technological preparation of production, the formation of knowledge, skills and abilities in the field of design of technological processes of various types of production. And also, the formation of future bachelors technological thinking on the basis of the economic nature of normalization and planning of production of materials and technological processes.

2. Objectives of the discipline:

- to the main objectives of the discipline include understanding the essence and features of the implementation of technological preparation of production (TPP), the ability to develop technological processes for different types of production;

-study of the theoretical basis for the development of progressive norms of material consumption, ways and methods of saving material resources.

3. place of the discipline in the structure of the educational program of higher education

The discipline belongs to the variative part of Block 1. Disciplines (modules) of the educational program and is a discipline of choice, deepening the mastering of the profile

4. A graduate who has mastered the discipline shall possess the following competencies:

PC-8 readiness to fulfill the basic requirements of office management in relation to records and protocols; to execute project and working technical documentation in accordance with normative documents

PC-16 - ability to use knowledge of traditional and new technological processes and operations, normative and methodological materials on technological preparation of production, quality, standardization and certification of products and processes with elements of economic analysis in production

PC-17 ability to use in professional activity the basics of technological process design, development of technological documentation, calculations and design of parts, including with the use of standard software tools.

5. The scope of the discipline (indicating the labor intensity of all types of academic
work)

Types of training work	Total labor intensity (in hours)		
	Total	Semester	
	ZE /hours	6	7
Total labor intensity of the discipline (module) in ZE/hour	7 ZE/252.	3 ZE/108.	4 ZE/144.
Contact work of students with the teacher by types of training sessions (classroom work), including:	***69***	**34,6**	**34,4**
Lectures	32	16	16
Laboratory work	-	-	-
Practical exercises	32	16	16
Coursework (consultation, defense)	2,3	2,3	-
Course project (consultations, defense)	-	-	-
Pre-exam counseling	2	-	2
Contact work at the intermediate control	0,7	0,3	0,4
Independent work of the student (extracurricular work), including:	***183***	*73,4*	***109,6***
Working through the training material (self-training)	115,7	39,7	76
Course project (preparation)	-	-	-
Coursework (preparation)	-	33,7	-
Preparation for interim certification	33,6	-	33,6
Intermediate certification		credit	examination

RPA developer: E.R. Galimov, Doctor of Technical Sciences, Professor, Department of MS&PS of KNITU-KAI, E.A. Solopova, E.S. Mukhametshina, Candidate of Technical Sciences, Associate Professor, Department of MS&PS.

B1.B.ДВ.09.02 "Organization of procurement production"

1.Purpose of studying the discipline (module)

The purpose of studying the discipline is to prepare a bachelor to develop measures for technological preparation of production, the formation of knowledge, skills and abilities in the field of design of technological processes of various types of production. And also, the formation of future bachelors technological thinking on the basis of the economic nature of normalization and planning of production of materials and technological processes.

2. Objectives of the discipline:

- to the main objectives of the discipline include understanding the essence and features of the implementation of technological preparation of production (TPP), the ability to develop technological processes for different types of production;

-study of the theoretical basis for the development of progressive norms of material consumption, ways and methods of saving material resources.

3. place of the discipline in the structure of the educational program of higher education

The discipline belongs to the variant part of Block 1. Disciplines (modules) of the educational program and is a discipline of choice, deepening the mastering of the profile

4. A graduate who has mastered the discipline shall possess the following competencies:

PC-8 readiness to fulfill the basic requirements of office management in relation to records and protocols; to execute project and working technical documentation in accordance with normative documents

PC-16 - ability to use knowledge of traditional and new technological processes and operations, normative and methodological materials on technological preparation of production, quality, standardization and certification of products and processes with elements of economic analysis in production

PC-17 ability to use in professional activities the basics of technological process design, development of technological documentation, calculations and design of parts, including the use of standard software tools.

5. The scope of the discipline (indicating the labor intensity of all types of academic
work)

Types of training work	**Total labor intensity (in hours)**		
	Total ZE /hours	**Semester**	
		6	**7**
Total labor intensity of the discipline (module) in ZE/hour	7 ZE/252.	3 ZE/108.	4 ZE/144.

Contact work of students with the teacher by types of training sessions (classroom work), including:	***69***	**34,6**	**34,4**
Lectures	32	16	16
Laboratory work	-	-	-
Practical exercises	32	16	16
Coursework (consultation, defense)	2,3	2,3	-
Course project (consultations, defense)	-	-	-
Pre-exam counseling	2	-	2
Contact work at the intermediate control	0,7	0,3	0,4
Independent work of the student (extracurricular work), including:	***183***	*73,4*	***109,6***
Working through the training material (self-training)	115,7	39,7	76
Course project (preparation)	-	-	-
Coursework (preparation)	-	33,7	-
Preparation for interim certification	33,6	-	33,6
Intermediate certification		credit	examination

RPA developer: E.R. Galimov, Doctor of Technical Sciences, Professor, Department of MS&PS of KNITU-KAI, E.A. Solopova, E.S. Mukhametshina, Candidate of Technical Sciences, Associate Professor, Department of MS&PS.

B1.B.ДB.10.01 "Theory and technology of processes of production, processing and recycling of materials and coating application"

1.Purpose of studying the discipline (module)

The main purpose of studying the discipline is the formation of future bachelors technological thinking on the basis of knowledge of technological processes of production of materials and coatings, as well as the basics of their design.

2. Objectives of the discipline:

- familiarization of students with multifunctional criterion systems of complex development of technological processes of their production and processing; management of structure and characteristics of specific groups of materials, semi-finished products and products;

-learning the practice of calculations of basic technological parameters and technological equipment for the processes of powder technologies, coatings, molding, plastics processing

- expansion, deepening and consolidation of theoretical knowledge and the combination of theory and practice is achieved in the performance of laboratory classes in the classrooms of the department, as well as during the period of practical training.

3. place of the discipline in the structure of the educational program of higher education

The discipline belongs to the variant part of Block 1. Disciplines (modules) of the educational program and is a discipline of choice, deepening the mastering of the profile

4. A graduate who has mastered the discipline shall possess the following competencies:

PC-9- readiness to participate in the development of technological processes of production and processing of coatings, materials and products from them, technological process control systems

PC-16- ability to use knowledge of traditional and new technological processes and operations, normative and methodological materials on technological preparation of production, quality, standardization and certification of products and processes with elements of economic analysis in production

5. The scope of the discipline (indicating the labor intensity of all types of academic
work)

Types of training work	Total labor intensity (in hours)		
	Total	Semester	
	ZE /hours	6	7
Total labor intensity of the discipline (module) in ZE/hour	7 ZE/252.	2 ZE/72.	5 ZE/180.
Contact work of students with the teacher by types of training sessions (classroom work), including:	101	48,3	52,7
Lectures	64	32	32
Laboratory work	32	16	16
Practical exercises	-	-	-
Coursework (consultation, defense)	2,3	-	2,3
Course project (consultations, defense)	-	-	-
Pre-exam counseling	2	-	2
Contact work at the intermediate control	0,7	0,3	0,4
Independent work of the student (extracurricular work), including:	151	23,7	127,3

Working through the training material (self-training), including.	83,7	23,7	60
Course project (preparation)	33,7	-	33,7
Coursework (preparation)	-	-	-
Preparation for interim certification	33,6	-	33,6
Intermediate certification		credit	examination

RPA developer: E.R. Galimov, Ph.D., Professor, Department of MS&PS of KNITU-KAI, T.A. Ilinkova, Ph. A.V. Chernoglazova, Candidate of Technical Sciences, Associate Professor, Department of MS&SPB

B1.V.DV.10.02 "Technological processes of materials production

1.Purpose of studying the discipline (module)

The main purpose of studying the discipline is the formation of future bachelors technological thinking on the basis of knowledge of technological processes of production of materials and coatings, as well as the basics of their design.

2. Objectives of the discipline:

- familiarization of students with multifunctional criterion systems of complex development of technological processes of their production and processing; management of structure and characteristics of specific groups of materials, semi-finished products and products;
-learning the practice of calculations of basic technological parameters and technological equipment for the processes of powder technologies, coatings, molding, plastics processing
- expansion, deepening and consolidation of theoretical knowledge and the combination of theory and practice is achieved in the performance of laboratory classes in the classrooms of the department, as well as during the period of practical training.

3. place of the discipline in the structure of the educational program of higher education

The discipline belongs to the variative part of Block 1. Disciplines (modules) of the educational program and is a discipline of choice, deepening the mastering of the profile

4. A graduate who has mastered the discipline shall possess the following competencies:

PC-9- readiness to participate in the development of technological processes of production and processing of coatings, materials and products from them, technological process control systems

PC-16- ability to use knowledge of traditional and new technological processes and operations, normative and methodological materials on technological preparation of production, quality, standardization and certification of products and processes with elements of economic analysis in production

5. The scope of the discipline (indicating the labor intensity of all types of academic work)

Types of training work	Total labor intensity (in hours)		
	Total ZE /hours	Semester	
		6	7
Total labor intensity of the discipline (module) in ZE/hour	7 ZE/252.	2 ZE/72.	5 ZE/180.
Contact work of students with the teacher by types of training sessions (classroom work), including:	101	48,3	52,7
Lectures	64	32	32
Laboratory work	32	16	16
Practical exercises	-	-	-
Coursework (consultation, defense)	2,3	-	2,3
Course project (consultations, defense)	-	-	-
Pre-exam counseling	2	-	2
Contact work at the intermediate control	0,7	0,3	0,4
Independent work of the student (extracurricular work), including:	151	23,7	127,3
Working through the training material (self-training),	83,7	23,7	60
Course project (preparation)	33,7	-	33,7
Coursework (preparation)	-	-	-
Preparation for interim certification	33,6	-	33,6
Intermediate certification		credit	examination

RPA developer: E.R. Galimov, Ph.D., Professor, Department of MS&PS of KNITU-KAI, T.A. Ilinkova, Ph. A.V. Chernoglazova, Candidate of Technical Sciences, Associate Professor, Department of MS&SPB

B1.B.01.01.01(U) "Familiarization practice"

1.Purpose of studying the discipline (module)

The purpose of the training practice is to familiarize students with the organizational structure of the university, the history of the graduating department, the content and organization of work carried out at the graduating department.

2. Objectives of the discipline:

- study of the organizational structure of the university and the management system operating in it; the history of the establishment and development of the MS and PB departments, the rules of training, the rights and obligations of students;
- familiarization with the program of the specialty being trained, lectures and projects;
- familiarization with the laboratories and equipment of the department and the university, the content of the main works and research carried out at the graduating department.
- acquiring skills of using modern information and communication technologies and global resources in research and computational-analytical activities in the field of materials science and materials technology;
- acquisition of skills of data collection, analysis of scientific and technical information, development and use of technical documentation, basic normative documents on intellectual property issues, preparation of documents for patenting, registration of know-how.

3. place of the discipline in the structure of the educational program of higher education

Training practice is a part of the variant part of Block B2.

Training practice is an integral part of the educational process and is a type of training sessions directly focused on the professional and practical training of bachelors.

Methods of training practice: stationary and/or on-site.

4. A graduate who has mastered the discipline shall possess the following competencies:

OK-2 ability to analyze the main stages and patterns of historical development of society to form a civic position
OK-7 ability to self-organization and self-education
OPK-1 ability to solve standard tasks of professional activity on the basis of information and bibliographic culture with the use of information and communication technologies and taking into account the basic requirements of information security.
PC-2- ability to collect data, study, analyze and summarize scientific and technical information on the subject of research, development and use of technical documentation, basic normative documents on intellectual property issues, preparation of documents for patenting, registration of know-how.

5. The scope of the discipline (indicating the labor intensity of all types of academic
work)

Types of training work	**Total labor intensity (in hours)**		
	Total ZE /hours	**Semester**	
		2	
Total labor intensity of the discipline (module) in ZE/hour	1ZE/36	1ZE/36	
Contact work of students with the teacher by types of training sessions (classroom work), including:	***12,3***	***12,3***	
Lectures	-	-	
Laboratory work	-	-	
Practical exercises	4	4	
Coursework (consultation, defense)	-	-	
Course project (consultations, defense)	-	-	
Pre-exam counseling	-	-	
Contact work at the intermediate control	0,3	0,3	
Independent work of the student (extracurricular work), including:	*23,7*	*23,7*	
Working through the training material (self-training)	-	-	
Course project (preparation)	-	-	
Coursework (preparation)	-	-	
Preparation for interim certification	-	-	
Intermediate certification		credit	

RPA Developer: E.R. Galimov, Doctor of Technical Sciences, Professor, Department of MS&PS of KNITU-KAI, P.B. Shibaev, Candidate of Technical Sciences, Associate Professor, Department of MS&PS.

B2.B.01.02(U) "Practice on obtaining primary professional skills, including primary skills and skills of research work".

1.Purpose of studying the discipline (module)

-reinforcement of theoretical and practical knowledge obtained during the study of basic disciplines;

-learning of techniques, methods and methods of identification, observation, measurement and control of parameters of production technological and other processes in accordance with the profile of training;

-learning of techniques, methods and ways of processing, presenting and interpreting the results of practical research;

-acquisition of practical skills in future professional activity or in its separate sections;

-development of creative approach to solving problems in the subject area, activization of cognitive activity of students

2. Objectives of the discipline:

mastering the methods of research and testing of metallic materials;

study of methods of sample preparation and analysis of the structure of materials;

study of methods of manufacturing various alloys and production of products from them;

acquisition of practical skills and abilities on the basis of individual assignments using modern engineering software tools;

mastering of techniques, methods and ways of processing, presentation and interpretation of the results of theoretical and practical research.

3. place of the discipline in the structure of the educational program of higher education

"Training practice for obtaining primary professional skills, including primary skills and skills of research work" belongs to the variable part of Block 2 "Practices".

Methods of training practice: stationary and/or on-site.

4. A graduate who has mastered the discipline shall possess the following competencies:

OK-2 ability to analyze the main stages and patterns of historical development of society to form a civic position
OK-7 ability to self-organization and self-education
OPK-1 ability to solve standard tasks of professional activity on the basis of information and bibliographic culture with the use of information and communication technologies and taking into account the basic requirements of information security.
PC-2- ability to collect data, study, analyze and summarize scientific and technical information on the subject of research, development and use of technical documentation, basic normative documents on intellectual property issues, preparation of documents for patenting, registration of know-how.

5. The scope of the discipline (indicating the labor intensity of all types of academic
work)

Types of training work	**Total labor intensity (in hours)**	
	Total hours (ZE)	**Semester**
		5
Total labor intensity of the discipline (module) in ZE/hour	3 ZE/108.	3 ZE/108.
Contact work of students with the teacher by types of training sessions (classroom work), including:	*32*	*32*
Contact work at the intermediate control	0,3	0,3
Independent work of the student (extracurricular work), including:	*75,7*	*75,7*
Working through the training material (self-training)	75,7	75,7
Preparation for interim certification		
Intermediate certification		Pass with a grade

RPA Developer: V.H. Abdullina, Candidate of Technical Sciences, Associate Professor of the MS&PS Department, A.V. Belyaev, Candidate of Technical Sciences, Associate Professor of the MS&PS Department.

B2.B.02.01(P) "Practice on obtaining professional skills and experience of professional activity"

1.Purpose of studying the discipline (module)

Formation of general and professional competencies, as well as the acquisition of necessary skills and experience of practical work by students in the specialty. During the internship the student gets acquainted with the organization of scientific, technical and production activities, laboratories, departments.

Consolidation of theoretical knowledge obtained during the study of basic and variable disciplines, mastery of the basics of materials research methods; mastering of methods and ways of designing blanks and technological processes of their manufacture; ability to apply materials for the manufacture of specific types of parts; acquisition of practical skills in the design of technological processes.

2. Objectives of the discipline:

- development and accumulation of special skills, study and participation in the development of organizational, methodological and regulatory documents for solving individual tasks at the place of practice;
-study of the organizational structure of the enterprise and the management system operating in it;
-acquaintance with the content of the main work and research performed at the enterprise or organization at the place of internship;
-Study of the structure, condition, behavior, and/or operation of specific technological processes;
-learning of techniques, methods and methods of identification, observation, measurement and control of parameters of production, technological and other processes in accordance with the profile of training;
-taking part in a specific production process or research;
-learning of techniques, methods and ways of processing, presenting and interpreting the results of practical research;
-acquisition of practical skills in future professional activity or in its separate sections,
- mastering the methods of research and testing of non-metallic, metallic and composite materials;
- mastering the development of drawings of workpieces and technological processes of their manufacture by types of production;
- mastering of methods of control and defectoscopy of parts for various purposes.

3. place of the discipline in the structure of the educational program of higher education

The discipline belongs to the part formed by participants of educational relations, Block 2. Disciplines (modules) of the educational program and is an elective discipline, deepening the mastering of the profile
Methods of conducting industrial practice: stationary and/or on-site.
4. A graduate who has mastered the discipline shall possess the following competencies:
OPK-3 readiness to apply fundamental mathematical, natural science and general engineering knowledge in professional activity.
OPK-5 ability to apply the principles of rational use of natural resources and environmental protection in practical activities
PC-8 readiness to fulfill the basic requirements of office management in relation to records and protocols; to execute design and working technical documentation in accordance with normative documents
PC-10 ability to assess the quality of materials in production conditions at the stage of pilot testing and implementation
PC-12 readiness to work on the equipment in accordance with the rules of safety, industrial sanitation, fire safety and labor protection norms
PC-16 ability to use knowledge of traditional and new technological processes and operations, normative and methodological materials on technological preparation of production, quality, standardization and certification of products and processes with elements of economic analysis in production

5. The scope of the discipline (indicating the labor intensity of all types of academic
work)

Types of training work	**Total labor intensity (in hours)**	
	Total ZE /hours	**Semester** **6**
Total labor intensity of the discipline (module) in ZE/hour	3 ZE/108.	3 ZE/108.
Contact work of students with the teacher by types of training sessions (classroom work), including:	***4,3***	***4,3***
Lectures	-	-
Laboratory work	-	-
Practical exercises	4	4
Coursework (consultation, defense)	-	-
Course project (consultations, defense)	-	-
Pre-exam counseling	-	-

Contact work at the intermediate control	0,3	0,3
Independent work of the student (extracurricular work), including:	***103,7***	***103,7***
Working through the training material (self-training)	103,7	103,7
Course project (preparation)	-	-
Coursework (preparation)	-	-
Preparation for interim certification	-	-
Intermediate certification		credit

Developer of RPA: S.V. Kuryntsev; Candidate of Economic Sciences, Associate Professor of the MS&PS Department, F.N. Kurtayeva; Candidate of Technical Sciences, Associate Professor of the MS&PS Department.

B2.B.02.02(P) Industrial practice - Research work

1.Purpose of studying the discipline (module)

Industrial practice - Research work is the final stage of training, and obtaining skills to work on research equipment is carried out after the students have mastered the program of theoretical and practical training.

The purpose of research work is to develop skills of independent solution of practical research problems, as well as mastering the functional responsibilities of officials in the profile of future work....

2. Objectives of the discipline:

- obtaining skills of work on research equipment used in the implementation of the diploma project (work);
- Know the standard methods of research and testing used to study the characteristics of materials, including certification methods;
- practical mastering of modern methods of scientific research, mathematical processing of results;
- use modern information and communication technologies, global information resources in research and calculation and analytical activities in the field of materials science and technology of materials;
- collect data, study, analyze and summarize scientific and technical information on the subject of research using technical documentation;
- using normative and methodological materials, prepare and execute technical assignments for measurements, tests, research and development work on the subject of the study.

3. place of the discipline in the structure of the educational program of higher education

Industrial Practice - Research Work is designed for fourth-year students and serves to form practical and consolidate theoretical knowledge

obtained during the study of basic and variable disciplines. Industrial Practice - Research Work is a part of Block 2 "Practices" of the variative part.
Methods of conducting industrial practice: stationary and/or on-site.
4. A graduate who has mastered the discipline shall possess the following competencies:
OPK-4 ability to combine theory and practice to solve engineering problems
PC-1 ability to use modern information and communication technologies, global information resources in research, calculation and analytical activities in the field of materials science and materials technology
PC-2 ability to collect data, study, analyze and summarize scientific and technical information on the subject of research, development and use of technical documentation, basic normative documents on intellectual property issues, preparation of documents for patenting, registration of know-how.
PC-4 ability to use in research and calculations knowledge of methods of research, analysis, diagnostics and modeling of properties of substances (materials), physical and chemical processes occurring in materials during their obtaining, processing and modification.
PC-5 readiness to perform complex research and tests in the study of materials and products, including standard and certification, their production, processing and modification processes
PC-6 ability to use in practice modern ideas about the influence of microstructure and nanostructure on the properties of materials, their interaction with the environment, fields, particles and radiation.
PC-13 Ability to use normative and methodological materials for preparation and execution of technical assignments for measurements, tests, research and development works

5. The scope of the discipline (indicating the labor intensity of all types of academic
work)

Types of training work	General labor intensity			Semester: **8**		
	in ZE	per hour	per week.	in ZE	per hour	per week.
Total labor intensity of the discipline (module)	**3**	**108**	**2**	**3**	**108**	**2**
Contact work of students with the teacher by types of		**4,3**			**4,3**	

training sessions (classroom work), including:						
Counseling		**4**			**4**	
Contact work at the intermediate control		**0,3**			**0,3**	
Student's independent work	**3**	**103,7**	**2**	**3**	**103,7**	**2**
Interim assessment:	credit					

Developer of RPA: S.V. Kuryntsev; Candidate of Economic Sciences, Associate Professor of the MS&PS Department, F.N. Kurtayeva; Candidate of Technical Sciences, Associate Professor of the MS&PS Department.

B2.B.02.03(P) Industrial practice - pre-diploma practice

1.Purpose of studying the discipline (module)

Pre-diploma practice is the final stage of training and is conducted after students have mastered the program of theoretical and practical training. The purpose of pre-diploma practice is to develop skills of independent solution of practical engineering tasks, as well as mastering the functional responsibilities of officials in the profile of future work.

2. Objectives of the discipline:

- elaboration of the topic of the final qualification work and selection of material for it;
- expansion and consolidation of knowledge on the profile of production;
- practical mastering of modern methods of scientific research, mathematical processing of results;
- fulfillment of special development to solve specific production tasks;
- expansion and consolidation of knowledge of economics and scientific organization of production;
- direct participation in the production and social life of the production team of the shop, department, laboratory....

3. place of the discipline in the structure of the educational program of higher education

Pre-diploma practice is designed for fourth-year students and serves to consolidate the theoretical knowledge obtained during the study of basic and variable disciplines and the formation of practical skills. Pre-diploma practice is a part of Block 2 "Practices" of the variative part.

Methods of conducting industrial practice: stationary and/or on-site.

4. A graduate who has mastered the discipline shall possess the following competencies:

OPK-5 ability to apply the principles of rational use of natural resources and environmental protection in practical activities
PC-3 readiness to use modeling methods in predicting and optimizing technological processes and properties of materials, standardization and certification of materials and processes
PC-9 readiness to participate in the development of technological processes of production and processing of coatings, materials and products from them, technological process control systems
PC-11 ability to apply knowledge of the main types of modern inorganic and organic materials, principles of materials selection for given operating conditions taking into account the requirements of manufacturability, cost-effectiveness, reliability and durability, environmental consequences of their application in the design of high-tech processes
PC-15 ability to ensure efficient, environmentally and technically safe production on the basis of mechanization and automation of production processes, selection and operation of equipment and tooling, methods and techniques of work organization PC-16 ability to use knowledge of traditional and new technological processes and operations, normative and methodological materials on technological preparation of production, quality, standardization and certification of products and processes with elements of economic analysis in production
PC-17 ability to use in professional activity the basics of technological process design, development of technological documentation, calculations and design of parts, including with the use of standard software tools.

5. The scope of the discipline (indicating the labor intensity of all types of academic
work)

Types of training work	General labor intensity			Semester: 8		
	in ZE	per hour	per week.	in ZE	per hour	per week.
Total labor intensity of the discipline (module)	14	504	8	14	504	8
Contact work of students with the teacher by types of training sessions (classroom work), including:		4,3			4,3	
Counseling		4			4	
Contact work at the intermediate control		0,3			0,3	

Student's independent work		499,7	8		499,7	8
Interim assessment:	credit					

Developer of RPA: S.V. Kuryntsev; Candidate of Economic Sciences, Associate Professor of the MS&PS Department, F.N. Kurtayeva; Candidate of Technical Sciences, Associate Professor of the MS&PS Department.

B3.B.01 Preparation for the defense procedure and defense of the final qualification work

1.Purpose of the GIA

The purpose of the state final attestation is to establish the conformity of the results of the students' mastering of the educational training program "Materials Science and Technology of New Materials" to the standard in the direction of training 22.03.01 "Materials Science and Technology of Materials".

2. GIA tasks:

The task of the GIA is to confirm the readiness to solve the following professional tasks, in accordance with the types of professional activities:

research and calculation and analytical activities:

collection of data on existing types and grades of materials, their structure and properties in relation to the solution of the set tasks using databases and literature sources;

participation in the work of a group of specialists in the performance of experiments and processing of their results on the creation, research and selection of materials, assessment of their technological and service qualities by complex analysis of their structure and properties, physical-mechanical, corrosion and other tests;

collection of scientific and technical information on the subject of experiments for the preparation of reviews, reports and scientific publications, participation in the preparation of reports on the completed assignment;

work with normative and technical documentation in the system of certification of materials and products, technological processes of their obtaining and processing, reporting documentation, records and protocols of the course and results of the experiment, documentation on safety and life safety;

participation in the work of a group of specialists in the development of technological processes of production, processing and modification of materials and coatings, parts and products, process control systems;

record keeping, execution of design and working technical

documentation, preparation of records and protocols at production sites;

fulfillment of the requirements of regulatory documentation in the development of design and technical documentation;

production and design and technological activities:

participation in obtaining and use (processing, operation and utilization) of materials for various purposes, design of high-tech processes at the stage of pilot testing and implementation;

participation in the organization of workplaces in the division, maintenance and diagnostics of measuring instruments and test equipment, control of compliance with quality requirements during measurements and tests, data processing;

participation in the development of technical assignments for measurements, tests, research and development work;

participation in the work on standardization, preparation and carrying out of certification of processes, equipment and materials, preparation of documents when creating a quality management system in the organization;

Designing high-tech processes as part of a primary design and technology or research unit;

development of design and working technical documentation;

3 Place of GIA in the structure of the Program of Graduate Studies

GIA: "Defense of work, including preparation for defense and defense procedure" is a part of the Basic part of Block 3.

4. As a result of mastering the Bachelor's degree program, the graduate shall have general cultural, general professional and professional competencies.

4.1A graduate who has mastered the Bachelor's degree program shall possess the following general cultural competencies:

ability to use the basics of philosophical knowledge to form a world outlook (OK-1);

ability to analyze the main stages and regularities of the historical development of society to form a civic position (OK-2);

ability to use the basics of economic knowledge in various spheres of activity (OK-3);

ability to use the basics of legal knowledge in various spheres of activity (OK-4);

ability to communicate orally and in writing in Russian and foreign languages to solve problems of interpersonal and intercultural interaction (OK-5);

ability to work in a team, tolerantly accepting social, ethnic, confessional and cultural differences (OK-6);
ability to self-organization and self-education (OK-7);
ability to use methods and means of physical culture to ensure full-fledged social and professional activity (OK-8);
readiness to use basic methods of protection of production personnel and population from possible consequences of accidents, catastrophes, natural disasters (OK-9).
4.2 A graduate who has mastered the Bachelor's degree program shall possess the following general professional competencies:
ability to solve standard tasks of professional activity on the basis of information and bibliographic culture with the use of information and communication technologies and taking into account the basic requirements of information security (OPK-1);
ability to use in professional activity the knowledge of approaches and methods of obtaining results in theoretical and experimental research (OPK-2);
readiness to apply fundamental mathematical, natural science and general engineering knowledge in professional activity (OPK-3);
ability to combine theory and practice to solve engineering problems (OPK-4);
ability to apply in practice the principles of rational use of natural resources and environmental protection (OPK-5).
4.3 A graduate who has mastered the Bachelor's degree program shall possess professional competencies corresponding to the type (types) of professional activity, which the Bachelor's degree program is focused on:
research and calculation and analytical activities:
ability to use modern information and communication technologies, global information resources in research and calculation and analytical activities in the field of materials science and technology of materials (PC-1);
ability to collect data, study, analyze and generalize scientific and technical information on the subject of research, development and use of technical documentation, basic normative documents on intellectual

property issues, preparation of documents for patenting, registration of know-how (PC-2);

readiness to use modeling methods in predicting and optimizing technological processes and properties of materials, standardization and certification of materials and processes (PC-3);

ability to use in research and calculations knowledge of methods of research, analysis, diagnostics and modeling of properties of substances (materials), physical and chemical processes occurring in materials during their obtaining, processing and modification (PC-4);

readiness to perform complex research and tests in the study of materials and products, including standard and certification, processes of their production, processing and modification (PC-5);

ability to use in practice modern ideas about the influence of micro- and nano-structure on the properties of materials, their interaction with the environment, fields, particles and radiation (PC-6);

ability to select and apply appropriate methods of modeling physical, chemical and technological processes (PC-7);

readiness to fulfill the basic requirements of office management in relation to records and protocols; to execute project and working technical documentation in accordance with normative documents (PC-8);

readiness to participate in the development of technological processes of production and processing of coatings, materials and products from them, technological process control systems (PC-9);

production and design and technological activities:

ability to assess the quality of materials in production conditions at the stage of pilot testing and implementation (PC-10);

ability to apply knowledge of the main types of modern inorganic and organic materials, principles of materials selection for given operating conditions taking into account the requirements of manufacturability, economy, reliability and durability, environmental consequences of their application in the design of high-tech processes (PC-11);

readiness to work on the equipment in accordance with the rules of safety, industrial sanitation, fire safety and labor protection norms (PC-12);

ability to use normative and methodological materials for preparation and execution of technical assignments for measurements, tests, research and development works (PC-13);
readiness to use technical means of measurement and control required for standardization and certification of materials and processes of their production, testing and production equipment (PC-14);
ability to ensure efficient, environmentally and technically safe production on the basis of mechanization and automation of production processes, selection and operation of equipment and tooling, methods and techniques of work organization (PC-15);
ability to use knowledge of traditional and new technological processes and operations, normative and methodological materials on technological preparation of production, quality, standardization and certification of products and processes with elements of economic analysis (PC-16);
ability to use in professional activity the basics of technological process design, development of technological documentation, calculations and design of parts, including the use of standard software (PC-17).

5. Scope of the GIA

Types of training work	General labor intensity			Semester: 8		
	in ZE	per hour	per week.	in ZE	per hour	per week.
Total labor intensity of the discipline (module)	6	216	4	6	216	4
Contact work of students with the teacher by types of training sessions (classroom work), including:		14,5			14,5	
GIA (execution and defense)		14,5			14,5	
Student's independent work		201,5	4		201,5	4
Final Attestation:	WRC defense					

RPA Developer: E.R. Galimov, Doctor of Technical Sciences, Professor, MS&PS Department, F.N. Kurtayeva; Candidate of Technical Sciences, Associate Professor, MS&PS Department.

B1.B.01.01.01(U) "Familiarization practice"

1.The purpose of studying the discipline (module): the discipline lays the foundation of legal knowledge, with the help of which the further activity of a specialist is carried out.

2. Objectives of the discipline:

1. Familiarization of students with a vast complex of knowledge about the state and law, with the system of knowledge about law;
2. formation of conceptual base in the field of jurisprudence;
3. Facilitating the assimilation of material that deals with the application and interpretation of law;
4. familiarization of students with the main branches of law regulating social relations;
5. promotion of mastering the basics of civil, criminal legislation, basics of labor, administrative and family law;
6. study of legal mechanism of realization of methods of state management of public life;
7. Study of conditions of realization of economic and business activity within the framework of legal activity;
8. training in lawful ways to protect their rights and legitimate interests.

3. place of the discipline in the structure of the educational program of higher education

The discipline belongs to the optional discipline, Block 1.

4. A graduate who has mastered the discipline shall possess the following competencies:

OK-4 - ability to use the basics of legal knowledge in various spheres of activity

5. The scope of the discipline (indicating the labor intensity of all types of academic work)

Types of training work	**Total labor intensity (in hours)**		
	Total ZE /hours	**Semester**	
		6	
Total labor intensity of the discipline (module) in ZE/hour	2 ZE/72.	2 ZE/72.	

Contact work of students with the teacher by types of training sessions (classroom work), including:	***34,3***	***34,3***	
Lectures	34	34	
Laboratory work	-	-	
Practical exercises	-	-	
Coursework (consultation, defense)	-	-	
Course project (consultations, defense)	-	-	
Pre-exam counseling	-	-	
Contact work at the intermediate control	0,3	0,3	
Independent work of the student (extracurricular work), including:	37,7	37,7	
Working through the training material (self-training)	37,7	37,7	
Course project (preparation)	-	-	
Coursework (preparation)	-	-	
Preparation for interim certification	-	-	
Intermediate certification		credit	

RPA Developer: E.R. Galimov, Doctor of Technical Sciences, Professor, Department of MS&PB, KNITU-KAI, V.I. Davlieva, Candidate of Legal Sciences, Associate Professor.

TOPIC 2. REVIEW OF ADVANCED RESEARCH IN MATERIALS SCIENCE

Objective. To find and analyze cutting edge research in materials science.

Assignment. Make a review of advanced research in the field of materials science with the help of open Internet resources. Prepare a presentation and a 5-minute paper.

TOPIC 3. OVERVIEW OF ADVANCED RESEARCH TRENDS IN MATERIALS SCIENCE AT LEADING UNIVERSITIES OF THE WORLD

Objective. To study and analyze advanced research in the field of materials science in the leading universities of the world.

Assignment. Make a review of advanced research in the field of materials science in the leading universities of the world with the help of open Internet resources. Prepare a presentation and a speech for 5 minutes.

QS World University Rankings by subject are based on academic reputation, employer reputation and research impact. https://www.topuniversities.com/university-rankings/university-subject-rankings/2021/materials-sciences

Institutions of higher learning:

1. Department of Materials Science & Metallurgy University of Cambridge https://www.msm.cam.ac.uk/research/research-groups
2. Nanyang Technological University, Singapore (NTU)
3. Department of Materials University of Oxfordhttp://www.materials.ox.ac.uk/
4. Department of Materials Science and Engineering University of California Berkeley http://www.mse.berkeley.edu/?fbclid=IwAR0q6yOVyboXn627ULrbAos13uLfycEIt6SD_qZ_Ii2FGbwzsqnRvq72Wd8

   et al.

THEME 4. ARTIFICIAL MATERIALS AND TECHNOLOGIES OF THEIR PRODUCTION BEFORE OUR ERA

Objective. To study and analyze man-made materials and technologies for their production before our era, as well as the contributions of individuals to their creation.

Assignment. Make an overview of artificial materials and technologies of their production in the period before our era, as well as the contribution of individuals in their creation, with the help of open Internet resources. Prepare a 5-minute presentation and performance.

THEME 5. MATERIALS AND TECHNOLOGIES OF THEIR PRODUCTION DURING THE RENAISSANCE (XIV-XVI CENTURIES). CONTRIBUTION OF **INDIVIDUAL PERSONALITIES TO THE DEVELOPMENT OF MATERIALS SCIENCE OF THIS PERIOD**

Objective. To study and analyze materials and technologies of their production in the Renaissance (XIV-XVI centuries). The contribution of individual personalities to the development of materials science of this period.

Assignment. To make an overview of existing materials and technologies of their production in the Renaissance (XIV-XVI centuries). Note the contribution of individual personalities in their creation and development of materials science of this period, with the help of open Internet resources. Prepare a presentation and performance for 5 minutes.

THEME 6. MATERIALS AND TECHNOLOGIES OF THEIR PRODUCTION IN THE NEW AGE (XVII-XIXTH CENTURIES). CONTRIBUTION OF INDIVIDUAL PERSONALITIES TO THE DEVELOPMENT OF MATERIALS SCIENCE OF THIS PERIOD

Objective. To study and analyze materials and technologies of their production in the New Age (XVII-XIX centuries). Contribution of individual personalities to the development of materials science of this period.

Assignment. To make an overview of existing materials and technologies of their production in the New Age (XVII-XIX centuries). To note the contribution of individual personalities in their creation and development of materials science of this period, with the help of open Internet resources. Prepare a presentation and performance for 5 minutes.

TOPIC 7. MATERIALS AND TECHNOLOGIES OF THEIR PRODUCTION IN THE MODERN ERA (XX-XXI CENTURIES). CONTRIBUTION OF INDIVIDUAL PERSONALITIES TO THE DEVELOPMENT OF MATERIALS SCIENCE OF THIS PERIOD

Objective. To study and analyze materials and technologies of their production in the Modern Age (XX-XXI centuries). Contribution of individual personalities to the development of materials science of this period.

Assignment. To make an overview of existing materials and technologies of their production in the Modern Age (XX-XXI centuries). To note the contribution of individual personalities in their creation and development of materials science of this period, with the help of open Internet resources. Prepare a presentation and performance for 5 minutes.

THEME 8. ACHIEVEMENTS OF DOMESTIC SCIENCE, TECHNOLOGY AND INDUSTRY IN THE FIELD OF MATERIALS. SCIENTISTS METALLURGISTS, METALLURGISTS AND MATERIAL SCIENTISTS OF RUSSIA

Objective. To study and analyze the achievements of domestic science, technology and industry in the field of materials. To learn about the contribution of scientists metallurgists, metallurgists and material scientists of Russia.

Assignment. To review the achievements of domestic science, technology and industry in the field of materials. Learn about the contribution of scientists metallurgists, metallurgists and material scientists of Russia, with the help of open Internet resources. Prepare a presentation and performance for 5 minutes.

LIST OF RECOMMENDED READING

1. Shibaev P.B. History of materials science: textbook for universities. / P.B. Shibaev et al. - Kazan: KSEU. 2014.- 257 c.

2. Introduction to Materials Science. https://openedu.ru/course/misis/MATSC1/

3. Processes of nanoparticles and nanomaterials production. ttps://openedu.ru/course/misis/NANOMAT/

yes

I want morebooks!

Buy your books fast and straightforward online - at one of world's fastest growing online book stores! Environmentally sound due to Print-on-Demand technologies.

Buy your books online at
www.morebooks.shop

Kaufen Sie Ihre Bücher schnell und unkompliziert online – auf einer der am schnellsten wachsenden Buchhandelsplattformen weltweit! Dank Print-On-Demand umwelt- und ressourcenschonend produzi ert.

Bücher schneller online kaufen
www.morebooks.shop

info@omniscriptum.com
www.omniscriptum.com

Printed by Books on Demand GmbH, Norderstedt / Germany